THE IMMORTALITY ENZYME

BIOMEDICAL BREAKTHROUGHS YOU CAN USE AGAINST AGING, CANCER & HEART DISEASE

by

PHILLIP MINTON, M.D.

WINNING Publications, Inc.
2372 Leibel St.
White Bear Lake, MN 55110

First Edition: December, 2000

Printed in the United States of America

ISBN: 1-884367-05-4

TABLE OF CONTENTS

SECTION ONE - IMMORTALITY

SECTION TWO - DETOXIFICATION

SECTION THREE - HORMONE REPLACEMENT THERAPY

SECTION FOUR - BODY AND MIND

SECTION FIVE - ALTERNATIVE CANCER TREATMENT

About the author:

Phillip Minton, M.D., M.Sc., has extensive experience and educational background in both alternative and conventional medicine. His keen interest in science, along with concern for the health of others, led him to a career in medicine. Graduating with honors from Southern Illinois University with a B.A. major in biology in 1979, he was allowed to undergo an accelerated three year medical school curriculum at the S.I.U. School of Medicine. He was granted his Doctor of Medicine (M.D.) degree three years later in 1982 from Southern Illinois University, Springfield, Illinois, U.S.A. Following internship in general medicine, he spent nearly a decade in a small rural Northern California town practicing general family medicine. During that time Dr. Minton cultivated his knowledge of alternative and complementary medicine including acupuncture and vitamin/mineral therapies. In 1994 he returned to school to bolster his education in the science and medical techniques of cancer and anti-aging medicine. This was accomplished by performing laboratory research on cancer biology and aging at Stanford University in Palo Alto California. Dr. Minton graduated from Stanford University in 1995 with a Masters Degree in Biology, major in molecular biology (M.Sc.).

Dr. Minton spent his spare time for several years studying the techniques of many of the world's most effective alternative clinics, with a special emphasis on alternative cancer clinics, interviewing the medical directors. This effort included visiting the clinics and even working for several of them. Doctor Minton now uses a combination of the best techniques gleaned from these alternative clinics plus the unique insights gained at Stanford University as the basis for his patient treatment regimens. While a graduate student at Stanford, Dr. Minton developed his uniquely powerful natural method of treating cancer. He calls this method R-A Therapy. The "R" is for "re-differentiation" (cancer cells reverting to a non-cancerous state), and the "A" stands for "apoptosis" (cancer cells deleting themselves from the body via "cellular suicide)." Doctor Minton finds that combining the tenets of R-A Therapy with the best of other, more widely-known, alternative cancer treatments (including European techniques) often produces dramatically effective results.

(775) 324-5700

521 Hammill Lane • Reno, NV 89511

SECTION ONE - IMMORTALITY

Chapter One
THE PROMISE OF ANTI-AGING MEDICINE

Longevity

Have you ever dreamed of physical immortality? Of living "forever young"? Perhaps the most compelling reason for the recent popularity of anti-aging medicine is the age-old search for immortality. Since the dawn of civilization, mankind has sought to live "younger." The search for youthful immortality led the ancient Egyptians to develop herbs and medicines to rejuvenate the elderly and to help the young retain their youthfulness. Ponce DeLeon sought the "Fountain of Youth" in the New World several hundred years ago, an effort which led to exploration of the Americas.

Modern scientific discoveries and the rediscovery, by the medical profession, of once forgotten herbs and natural therapies is now bringing the dream of "youthful immortality" closer to realization than ever before. Therapies available today offer rejuvenation through replacement of youthful hormonal functioning, removal of toxic and waste materials from the body, natural stimulation of the organs via vitamins, minerals, herbs, and advanced medications. Nootropic medicines are a new class of pharmaceuticals, which stimulate and rejuvenate the functioning of the brain, thereby improving memory, alertness, and intelligence. They also fight senile dementia. Novel treatments and preventive regimens for cancer, arthritis, and heart disorders are now either immediately available or are soon to arrive. Today, the promise of living a long life, free of degenerative diseases such as cancer and heart disease is becoming a reality!

One of the most exciting areas of biomedical science, an area known as "molecular biology," studies the genes and other tiny components of cells. Cells are the smallest known components of living organisms. Billions of cells together make up the human body. Recently discovered genes are now believed to control important components of the aging process. One of them, the telomerase gene, may actually be the "immortality gene" which allows certain microorganisms and cells to live indefinitely. Several biotechnology companies are currently investing millions of dollars developing ways to confer this immortality gene to humans, and to delete it from cells which are made dangerous because of their "immortality." For the first time in the history of mankind, we are living in an age in which we can slow the aging process and in which we may soon be able to reverse it altogether. This book discusses many of these leading edge "biotherapies" for aging and aging-related degenerative diseases such as cancer and heart disease.

Energy, Vigor and Vitality

Lack of energy is one of the most common health complaints in modern society. With age, energy levels and vigor decline. Aging related decreases in energy levels commonly begin during the third decade of life. Along with this "low energy condition," the perception of vitality and the "zest for life" also diminishes. Even mental energy wanes. Depression is a medical diagnosis that connotes a substandard "mental energy" level. Diagnosed cases of depression become more and more common as people age. One of the most commonly treated medical conditions among the elderly is depression.

How can we deal with the loss of energy, vigor, and vitality that occurs as we age? Certainly, low energy levels can stem from various sources in each individual. However, many common factors have been found to explain the loss of energy that occurs with the aging process. Lack of hormones can lead to a tired condition. Low thyroid and adrenal hormone levels are also common problems. Statistically, hypothyroidism (low thyroid hormone functioning) rates increase dramatically in older persons. Similarly, low levels of melatonin hormone can lead to disrupted and restless sleep, causing fatigue for the rest of the day. Poor nutrition is also a common cause of tiredness and lethargy in older people. With age, our ability to digest, absorb, and assimilate nutrients from food decreases, and it becomes difficult to utilize nutrients from vitamins and other nutritional supplements. Dysfunctions of digestion and utilization of nutrients, as well as other factors impairing energy production in the body can be dealt with through many of the powerful alternative medical techniques described throughout this book.

Sexual Functioning and Sexual Energy

Complaints about diminishing sexual ability and poor sexual functioning become more and more common with age. Restoration of declining sexual ability ranks among the top of those problems listed by people interested in anti-aging medical therapies. As hormone levels decline, and general health and vigor abate, sexual functioning and sexual energy also decline. Both sexes complain of 1) lost libido (sexual desire) and 2) decreased general energy levels which impair the enjoyment of sexual relations. Low libido, a direct result of low sex hormone levels, can be restored in many people by augmenting their sex hormones to more youthful levels. Poor levels of general energy can be improved and leads to better sexual performance and enjoyment.

Physical deterioration of the sex organs in both men and women can also be dealt with by several different therapies. Once again, hormonal therapy can rejuvenate and repair some of the deterioration. Specific medications, herbs, vitamins, minerals, and homeopathics can also be of great benefit. Male impotence, perhaps the most difficult sexual complaint to treat, can be improved for many sufferers. Potent herbs such as yohimbine and recently developed medications can be used if other anti-aging therapies fail to restore male potency. Female complaints such as vaginal dryness, pain on intercourse, burning on urination, and thinning and fragility of the vaginal lining, can be dealt with directly if they remain problematic after other anti-aging therapies and practices have been instituted.

Cosmetic Anti-Aging and Aging of the Skin

The skin is the most noticeable "organ" of the human body and provides several essential functions to the rest of the body. It serves as a barrier to infection and contamination from the outside world. It is an excretory organ, which like the kidneys, cleanses the body of waste products and toxins. The skin also helps to maintain the body's internal environment including temperature, fluid status, and chemistry.

Our skin is exposed to potent aging forces found in the external environment, such as sunlight, and is therefore especially prone to deterioration with age. Aging changes of the face are very noticeable, and lead to the expenditure of countless millions of dollars as we attempt to hide wrinkles and blemishes. It is now possible to actually slow or diminish the signs of aging of the skin, rather than simply covering them with makeup. Topical vitamins and nutrients can strengthen the skin tissues, decrease the rate of oxidative free radical damage, and slow the production of cross-linking. Cross-linking (abnormal chemical bonding) of the skin and subdermal (under the skin) tissues is thought to cause wrinkling, blemishes, and thinning of the skin. Alpha hydroxy acids are natural plant derived substances which can help to rejuvenate the skin by removing glue-like materials which stiffen and wrinkle the skin. Retin A is a medicine based on vitamin A. It delivers a vitamin-like substance to the subdermal tissues. The use of Retin A thereby results in less wrinkling, less blemishes, and a more youthful appearance of the skin.

Dermal peels and dermabrasion are two different modalities used to remove dead skin cells, thereby stimulating the production of new, healthier, and younger looking skin. Dermal peels achieve this removal of dead skin layers by chemically removing them from the skin surface. Dermabrasion is done by a physician who utilizes medical equipment to "abrade" or "sand off" the dead layers of skin cells. Dermabrasion also results in stimulating the skin to produce new, younger looking cells.

Systemic anti-aging therapies such as hormone replacement therapy can also decrease thinning, sagging, and wrinkling of the skin. A so-called "medical facelift" is sometimes attributed to the actions of human growth hormone replacement therapy. Strengthening of the skin and subdermal muscles and connective tissue structures by systemic growth hormone replacement has been known to dramatically lessen wrinkling, sagging and bagginess of the facial skin.

Dementia and Cognitive Enhancement

A steady decline in mental functioning and the ability to think and remember occurs as a normal aspect of human aging. Cognitive functions are those which involve our intelligence, memory, and ability to think or reason. A severe deficit in cognitive mental functioning is termed "dementia." General anti-aging procedures can improve cognitive function. More specific therapies, such as certain anti-aging herbs and medications, are even more effective. Nootropic drugs can be used to rejuvenate the memory and cognitive function in those suffering from dementia. They can also be employed to preserve the cognitive abilities of those who have not yet noticed diminished mental functioning.

Alzheimer's disease, a form of severely debilitating dementia, affects many older individuals. Although there is presently no consensus on the cause of this debilitating disease, it is known that abnormal aluminum deposits are associated with the damaged brain cells of sufferers. Detoxification may help to cleanse the brain of abnormal deposits of metals such as aluminum, mercury, lead and calcium. These metals are thought to play a role in several of the most common types of both senile and pre-senile dementia.

Degenerative Diseases

The many individual processes which together cause aging result in a gradual deterioration of the body and its physiological processes. The aging process is related to many diseases that more commonly afflict the elderly, or are more likely to occur as we grow older. These are the aging-related degenerative diseases and include many autoimmune diseases, diseases of the heart and blood vessels (cardiovascular disease), cancer, and arthritis.

Why are some diseases so prevalent as we grow old, but are fairly rare in youth? It may be that as bodily functions decline with age, these degenerative "aging-related diseases" are able to take hold. Just as it is difficult for weeds to germinate and develop in a well-manicured garden, these aging-related degenerative diseases probably are simply unable to "take root" and "germinate" in the healthier environment of a young body.

Several specific degenerative changes of the aging process may together allow an unhealthy environment suitable for the development of such aging-related diseases as heart and blood vessel disease, cancer and arthritis. These degenerative changes include:

(1) Deterioration of cellular and tissue functioning due to accumulated wastes, poor cell reproduction and hormonal imbalances,

(2) Deterioration of immune system functioning so that abnormal cells are left to proliferate, infective organisms are better able to cause damage, and dysfunctional immune cells become "confused" and attack our own healthy tissues, and

(3) Genetic abnormalities accumulate and directly lead to the production of cancerous and other genetically defective cells.

Therapies that slow or reverse the aging process conceivably have dramatic beneficial effects on these aging-related degenerative diseases. Following is a discussion of several of these diseases that may theoretically be benefited by anti-aging therapies.

Cardiovascular Disease

The bodily processes that comprise human aging may have a great impact on cardiovascular disease. Cardiovascular disease is a leading cause of death. It encompasses ailments such as heart attacks, strokes, phlebitis, aneurysms, and intermittent claudication from blocked arteries. All are very common afflictions in modern society. Researchers at the world's leading institutions are experimenting with newly discovered natural substances which control both aging and disease processes of the heart and blood vessels. These substances (PDGF, EDGF, Telomerase, etc.) slow the aging of the cardiovascular tissues and thereby decrease the onset and severity of many of the common diseases of the heart and blood vessels. It appears that as the tissues of the heart and blood vessels age, they become prone to the development of degenerative processes such as atherosclerosis (hardening of the arteries), heart attacks, strokes, blockages and inflammation of blood vessels in the legs, and heart rhythm disturbances such as "atrial fibrillation."

Incidences of these cardiovascular diseases may lessen or decline by many anti-aging therapies. Specific examples include the following:

(1) Minerals, vitamins, and digestive enzymes may benefit certain diseases. Calcium and potassium may decrease blood pressure; magnesium may decrease blood pressure and diminish arrhythmia. Vitamin C, vitamin E and B vitamins all have been shown to benefit cardiovascular disease. Digestive enzymes may decrease blood cholesterol levels in some people,

(2) Detoxification may benefit hypertension, arrhythmia, arterial blockages, and increase blood flow to essential organs such as the brain and heart muscle,

(3) Hormone Replacement Therapy (HRT) with female hormones may reduce the incidence of cardiovascular disease in women. Human growth hormone therapy may reduce cardiovascular disease in both sexes. Recent studies have suggested that human growth hormones may be especially effective in treating certain types of "cardiomyopathy," which is the abnormal enlargement and weakening of the heart muscle, and

(4) Exercise, herbs and homeopathic remedies may also decrease the severity of heart disease or delay its onset.

Cancer

Probably the most feared of the aging-related degenerative diseases, cancer seems to have touched everyone either personally or through family and friends. It is well known that cancer incidence increases as we grow older. Why is this? It may be that many of the same processes that comprise aging also lead to cancer.

Cancer develops in two stages. The first stage is called "initiation," when a normal cell becomes "transformed" into a cancer cell. Some of the aging-related changes in cells that may make this "transformation" more likely include, 1) DNA and chromosomal deterioration, 2) collections of cellular waste and toxins, 3) lack of oxygen due to poor circulation, and 4) deteriorated external (hormonal) controls over cell reproduction and function. The next stage is "regression." It is not easy for a single cancer cell to survive and grow into a tumor if the body has a healthy internal environment. However, in an unhealthy bodily environment, the "transformed" cancer cell has a better chance to survive and then progress from a single cell to a tumor and then to metastasize into remote parts of the body. Antioxidants, healthy dietary fat ratios, and other dietary components of an anti-aging program may conceivably lessen the chance that a cancer cell can survive and progress into life threatening stages of disease. A healthy, "youthful" immune system may also lessen the likelihood of cancer cell survival by seeking it out and destroying it before it can develop and spread. Healthy blood circulation and youthful hormone levels may also strengthen the body to fight transformed cancer cells while they are small, weak, and vulnerable.

Chapter Two
THE IMMORTALITY ENZYME

Why We Age

Perhaps the most critical element in the search for workable life extension techniques is to understand the actual physical mechanisms in the body that cause aging and related degenerative diseases to occur. One beneficial way of attempting to understand why human beings are not immortal is to examine those organisms and cells of the body that are immortal. Certain organisms, particularly the very ancient, microscopic, one-celled organisms that live within the soil of the earth are, in general, immortal. By immortal, I mean that these cells do not seem to have any set life span. These microscopic cells generally reproduce by simply increasing in size until they have reached a certain volume at which they divide into two identical new cells. Therefore, the two new cells are simply direct continuations of the previous single, old cell. In so doing, these cells are essentially immortal. They will never die by the aging process. Their demise occurs only if they are killed by external forces or ingested by other organisms.

Likewise, there is one cell type within the body that is also immortal. We can broadly classify all the cells in our body as either somatic (the cells which comprise all of the tissues and organs except the reproductive glands) and the germinal cells (those which form the egg and sperm). The so-called germinal cells do not undergo aging and theoretically have an indefinite life span. The germinal cells are the reproductive cells (sperm and egg) that form a new human organism when combined through sexual relations. Therefore, the sperm cells in the male and the ovum cells or egg cells in the female are the only known cells in the human body that do not age and are essentially immortal. The rest of the cells of the human body, as far as we know, are mortal and have definite, set average life spans. In order to better understand aging, we will examine the differences between the germinal cells of the human body and the somatic cells.

The germinal cells of the body seem to have at least two major protections from aging. These protections include, 1) the presence of the enzyme telomerase that protects the chromosomes of the germinal cells from slowly being eroded over time, and 2) special protections from oxidative damage and from chemical damage. These two special protections are in contrast to the rest of the cells of the body. The remaining somatic cells of the body that make up ninety-nine percent of our physical body have a less intense free radical protection and chemical protection system. Most significantly, they appear to lack the telomerase enzyme that protects chromosomes from gradual degradation.

A further way to look at the process of human aging is to consider the stages of development of the human being. At conception, the one single cell, which has resulted from the union of the egg and sperm cells, begins dividing. As it divides over and over again, the resultant daughter cells gradually change and specialize into cells that become the tissues and organs that make up the growing fetus. At term, it has reached the developmental stage of a human infant. When the infant is born there is still a tremendous amount of growth and development that will occur until it becomes a reproductively able adult. The brain development, the spinal cord development and peripheral nerve development still have a great deal more to go before the infant can develop normal human functions including the full range of human thought processes. It is interesting to note that even in a newborn infant the immortal cells of the body, the sex cells, are essentially already formed and at their mature stage. They are not yet capable of fulfilling their roles in sexual reproduction because the rest of the cells of the body that support them are still immature and incapable of sexual reproduction. However, these germinal cells themselves are fully mature and are already protected from aging and many kinds of degradation. As the child develops through infancy, childhood, and adolescence it will later become a fully mature young adult capable of reproduction and care of the offspring.

In the early to mid twenties the first signs of human aging can start to be seen. On a time line continuum, it appears that human beings are pre-programmed at the moment of conception to develop into young adults with the ability to sexually reproduce and care for offspring. However, past the age of sexual maturity, there seems to be no definite program for proceeding. This is when aging begins. For some people in their early twenties, hair begins to turn white, balding begins to occur and the most delicate area of the skin of the body--that of the face--begins to show the very first signs of wrinkling and deterioration. It appears that the human organism has been pre-programmed at the time of conception to develop into the state of adulthood equivalent to a human in their early twenties. However, there is no evidence of any sort of program to take the human organism past that point. By the time the body has reached the twenties the organ systems are at their peak, but they are already beginning to turn off and in some cases shut down altogether.

An example of this is the thymus gland of the immune system. The thymus gland is essential for the development of the immune system in infants and children. It remains a large viable organ until around the time of puberty. During this time it begins to shrink and atrophy. This shrinkage is known as "involution of the thymus" and is substantially complete by the time of young adulthood. This indicates that in young adults the immune system has come to its peak and is ready to begin a slow and gradual decline, as are almost all of the organ systems of our body. The only cells of the body that appear to remain fairly well intact are the germinal cells of the body. In women, even these can show some signs of deterioration with advancing decades. After young adulthood is reached, the human organism gradually deteriorates until death occurs simply from old age if not by some age-related degenerative disease. Given the examination of this time line of human development and the gradual deterioration after sexual maturity, it seems likely that humans have a definite program to reach sexual maturity and to be strong and viable during the years of bearing and raising children. After that point has been reached, there appears to be no definite plan or program.

Given these facts, it seems reasonable to conclude that human aging may be a combination of factors which are all based around the lack of any program defining what the body is to do after it has reached sexual maturity. If the body has no pre-programmed instructions for what to do after sexual maturity is reached, it seems reasonable that the body may simply begin to slowly deteriorate from the point it is pre-programmed to reach. This deterioration stems from two major sources. These two major sources are, 1) "random damage" due to free radicals, chemicals, radiation and other deteriorating forces with which the body comes into contact, and 2) "preprogrammed obsolescence" caused, to a great extent, by the immortality enzyme known as "telomerase," and its relatives.

The Immortality Enzyme

"Pre-programmed obsolescence" describes the known propensities of many of the somatic body cells to have a limited life span due to a limited number of times they can divide and reproduce themselves. The primary molecular basis for pre-programmed obsolescence appears to be the lack of the enzyme "telomerase" within the somatic cells of our body.

Our cells contain chromosomes, in which are found the "genes" that are essential for the control of each cell and the organism which the cells combine to create. If telomerase is present, it can protect the genes and chromosomes. Without telomerase to protect the chromosomes, a portion of the chromosomes are lost each time they divide and reproduce. After many divisions, the body's cells will have lost such a significant amount of chromosomal material that they are no longer able to reproduce themselves. At this point the cells of the various organs and tissues of the body are not able to rejuvenate themselves through the reproduction of new cells and are in even greater danger of destruction due to random damage from free radicals and other deteriorating influences.

It may also be possible that the organs and tissues of our body are designed to constantly regenerate themselves through the formation of new cells and the replacement of the old cells by the new cells. This process may eventually be halted in cells lacking telomerase if they reach a critical threshold of chromosome loss and the resultant inability to reproduce themselves. This leads to dysfunctional tissues and organs. These dysfunctional tissues and organs may form the basis of the process we know of as "senescence" or "old age." The lack of a program for the body to follow after young adulthood leads to aging as the gradual accumulation of random damage to the tissues of the body and pre-programmed obsolescence of the cells of the body occurs.

Telomere Timeclock

A "timing" mechanism, controlling aging, may exist in each cell. As previously mentioned, the pre-programmed obsolescence theory is based on the recently discovered telomerase enzyme. The interaction of the telomerase enzyme, or the lack of the telomerase enzyme, with the gradual deterioration of the chromosomes of the body as the cells reproduce themselves sets up what appears to be a timing mechanism. The so called "telomere time clock" may be one of the most important factors in setting the maximum life span for human beings.

How does this "telomere timeclock" work? The body's cells can be thought of as being tiny, microscopic, fluid-filled bags. The outer portion of the bag is a lining known as the cell membrane. The cell membrane, which acts in a similar fashion to the skin of our body, is the outer edge of each cell. It protects the cell from the external environment and encloses all of the living, functioning components of the cell within the cell's interior. Inside this cell membrane is the main part of the cell known as the cytoplasm. The cytoplasm is a fluid area which contains tiny living components of the cell which produce energy for the cell and give the cell its ability to perform all of the functions required of it in life. In the center of the cell, surrounded by the cytoplasm, is the control mechanism of the cell. This control mechanism area is known as the nucleus of the cell. The nucleus contains the DNA and the "genes" in the form of chromosomes. Not only are chromosomes the blueprint of the original cell, they also contain the blueprint for the functions that the cell is to undergo during its life span.

At fairly frequent intervals, depending on the type of cell involved, each cell will reproduce itself by forming two new "daughter" cells. However, when our cells divide into two daughter cells, each of these will contain slightly less of the original chromosome material than the progenitor cell contained. This is in contrast to the immortal one-celled organisms that do not lose any of their chromosome material at each cell division. Therefore, as the cells of the body gradually divide and re-divide, they will slowly lose their chromosome material until they are unable to reproduce.

The actual mechanism of the cell's inability to reproduce itself is that the ends of the chromosomes are gradually lost at each cell division. The chromosomal end-pieces, known as telomeres, are crucial because they are the starting point for enzymes that "read" the information contained in each chromosome. The ends of the chromosomes contain the portions of the chromosome material to which the enzymes that read the chromosomes attach themselves and begin the reading process. Enzymes crucial to life processes read the chromosomes only if those enzymes can attach to the end segment. Loss of the end segments therefore does not allow attachment and reading by the enzymes. If they can attach and read the end segment, the enzymes produce proteins and other materials that form the cells, and the material necessary for the cells to perform their living functions. Once these end portions of the chromosomes have reached a critically short length due to gradual loss at each cell division, the enzymes (known to science as transcription enzymes) are no longer able to attach and read the chromosome. At that point, two significant functions are lost. First, the ability of the chromosome to completely reproduce itself and form a daughter cell is lost. Secondly, significant defects in the ability of the enzymes to read the chromosomes for production of the materials needed for the cells to perform its living functions also occur. The combination of these two defects then makes the cells old, senescent, and within a relatively short period of time, death will occur.

The end segments of the chromosomes gradually deteriorate due to the lack of the telomerase enzyme which, if present, would have protected and preserved the telomeres (ends of the chromosomes). These chromosomal ends that gradually deteriorate in the cells of our body are known as "telomeres." The normal somatic cells of our body do not contain the telomerase enzyme. Therefore, since the telomerase enzyme is not present, it does not preserve the telomere ends of our body cells from gradual deterioration at each reproduction. This lack of telomerase, therefore, appears to be the essential basic reason for the pre-programmed obsolescence based upon the telomere time clock. Just like a grandfather clock slowly ticking away the minutes and the hours, as each cell division occurs, a small portion of the telomere end of the chromosome is lost. The telomere time clock is slowly ticking and ticking down the time until a critical amount of chromosomal telomere deterioration will have occurred whereby the cell is no longer able to reproduce or complete its normal functions of life.

This is in contrast to the immortal cells of our body, known as the germinal cells. Germinal cells, the egg and sperm cells, do have the telomerase enzyme. Therefore, no matter how many thousands or millions of times these cells divide throughout the eons of time on planet earth, they maintain their entire chromosomal length. Their telomeres do not shorten with time and are maintained at exactly the same length, as they were thousands or perhaps millions of years ago. Therefore, there is no pre-programmed obsolescence and no telomere time clock for these germinal cells of our body. They are essentially immortal. This is also true for the immortal one-celled organisms that live on the earth. These small one-celled microscopic organisms also contain the enzyme telomerase and therefore their telomere chromosome ends are protected from gradual deterioration, enabling these organisms to be essentially immortal.

Another very important class of cell that may develop within our body is also "immortal." These are cancer cells. A prime reason cancer cells are so deadly and can grow to a point that they will kill the entire organism is that they are immortal. Cancer cells are normal cells of the body that have changed in significant ways. The changes include the development in the cancer cells of mechanisms to produce the enzyme telomerase. Cancer cells producing telomerase are able to protect their cancerous chromosomes from gradual degradation. This means cancer cells can theoretically divide forever without ever coming to a point at which they are unable to divide, essentially conferring immortality upon them.

Cancer cells are immortal as long as they live within an environment in which they receive sustenance such as oxygen and nutrients. Until the entire organism in which they live dies, cancer cells will survive indefinitely. If removed from the original animal host these cancer cells, if given the nutrients they need, can live in the laboratory forever. Once again, this is in great contrast to the normal cells of our body, which are unable to do this because they lack the enzyme telomerase. Telomerase and the telomere time clock therefore appear to be the basis for both the pre-programmed obsolescence and aging of the human body and for the dangerous nature of cancer.

Telomerase was discovered just a few years ago. At the time of its discovery the incredible implications of what might be possible by manipulation of telomerase within human cells became readily apparent. It seems obvious that if the telomerase function can be influenced or changed in certain cell types of our body, dramatic medical progress can be made. If the telomerase present within cancer cells, which make them immortal, can be turned off, we may have for the first time a dramatic cure for many, if not all types, of cancers. Additionally, if telomerase activity can be restored in the healthy cells of the body they may theoretically be made immortal. In other words, the pre-programmed obsolescence portion of the multi-factorial process, known as aging, would be eliminated. If the cells of the body do not have to be subjected to the telomerase time clock then the potential human life span will become indefinite. The only limitations to this immortality would be accumulated random damage to the cells and tissues of the body from factors that are discussed next.

These factors are the random damage portion of the equation, which consists of several different elements. These elements are, 1) the gradual buildup of intra-cellular waste products, 2) deterioration from free radical oxidative damages, damage from radiation, chemical exposure, x-rays, and other external factors that affect the cells, and 3) the loss of homeostatic control at both the level of the cell itself and of the organism as a whole. The potential ramifications of gradual build up of waste products within the cells of the body will be discussed first.

As the body digests food and uses it through metabolism to feed the cells, waste products are produced. The largest portions of the waste products are eliminated through the body by way of the feces and the urine. However, there is some tiny percentage of the metabolic waste products that the cells are not able to eliminate from themselves. This tiny percentage of the waste products gradually accumulates within the cells and may, upon reaching a certain mass, interfere with the life functions of the cells. In this manner, the cells may slowly pollute themselves to death. The accumulation of waste products within the cells seems to occur primarily as the cells gradually age. Aging via the telomerase time clock may influence the ability of the cells to rid themselves of their waste products. Older cells seem to accumulate waste products at a greater rate than younger cells and may therefore slowly pollute themselves or poison themselves to death, whereas younger cells are more readily able to rid themselves of these destructive poisons.

One of the ways waste products are accumulated within the cells is in the form of lipofuscin granules. Lipofuscin granuals appear to be collections of waste products that are walled off by the cells within small compartments that exist inside the cytoplasm of cells. If the lipofuscin granules become large enough, they become visible to the naked eye. One of the places in which we can see lipofuscin-laden cells is in the age spots of our skin. Age spots are collections of cells that have large enough collections of lipofuscin that the pigment is easily seen on the surface of the skin. Evaluation of the brains of autopsied, aged individuals reveals that lipofuscin also accumulates within the brain cells and is apparent as dark areas in the tissues of the brain. It seems plausible that when lipofusion and other waste products reach a certain percentage of the cell's volume, they interfere with the cell's physiological functions and cause it to deteriorate and later die. Once again, this accumulation of waste products and poisoning of the cells by waste products may be intimately related to the main mechanisms of aging such as the telomere time clock and oxidative damage of the cells.

Gradual damage of the cells from various factors such as free radical damage may lead to deterioration of cells' ability to undergo and maintain life functions. Maintaining these life functions is known as "homeostasis." Homeostatic control of the cells allows them to maintain a metabolic equilibrium and function in the way they were designed. Their function is to maintain themselves and to act with other cells to form the working tissues and organs of our body.

One of the ways that homeostatic control may be lost on the level of the cell is through free radical damage and other types of external damage to the cells. Free radicals are reactive chemical species that interact with the components of our cells causing the various components to deteriorate. We might think of free radical damage as being the rust of our cells. As a piece of metal is oxidized the metal deteriorates into rust. This rust gradually flakes off, allowing more oxygen to oxidize new metal surfaces. New portions of the metal form rust until the entire piece of metal has been oxidized into rust. In like manner, oxygen, although essential for the metabolism of our cells, has a destructive side effect. This occurs when it combines with certain chemicals to form oxidated free radicals. These free radicals act to oxidize components of the cells causing them to deteriorate. Deterioration must be constantly protected against and repaired when it occurs. At birth, the body has some very powerful means of fighting the oxidating free radicals before they are able to interact with our cells and oxidize them.

One of the means by which our body can fight the oxidative free radicals is through the enzyme called superoxide dismutase. Superoxide dismutase is an abundant enzyme found throughout the entire body. Superoxide dismutase and a related enzyme called catalase, act to search out and neutralize oxidative free radicals wherever they develop within the body. Along with the enzymes, superoxide dismutase and catalase, the body has many other free radical fighting constituents.

Vitamin C is a very important free radical fighting substance found within the body. When vitamin C, also called ascorbic acid, is ingested, it can interact with many different types of oxidative free radicals, particularly those which are found within the watery components of the cells and the human body. Vitamin C is a powerful anti-oxidant that is crucial to good health. Many animals actually produce it within their bodies, but humans must ingest it daily. Many species of animals produce C in such large amounts that they never need to ingest foods containing it.

Vitamin E is another powerful anti-oxidant substance found in our food. It is most effective in quenching oxidative free radicals within the fatty components of our body and within the fatty components of the cells of our body. Vitamin E is a lipophilic, "fat-loving" substance and can readily dissolve itself into the fatty and lipid components of the cells. It is there that it searches out oxidative free radicals and quenches them before they are able to damage the cell membranes. In addition to these protective mechanisms against oxidized free radicals, the cells and tissues of the human body also have a mechanism that repairs the damage produced by those oxidated free radicals that get by the defense lines and damage the tissues and cells. These repair mechanisms are at a very high level of efficiency in a young body but as time passes they become weak and perform poorly.

Scientists are presently seeking to discover just exactly what this rate of oxidative damage and damage from external forces is, and at what point in the human life span its effects are greatest. It seems likely that the effects of oxidative free radicals and other external oxidative forces such as X-rays, radiation and chemicals occur primarily as the body weakens due to aging. In other words, oxidated free radical damage may simply be a magnifying factor within the primary aging process.

The primary aging process may be based upon the telomere time clock. In such a scenario, the human body is pre-programmed to develop into the stage of young adulthood, but thereafter does not have a program for continuing development or for maintenance of the youthful form. After reaching adulthood, various cells of the tissues of our body reach critically short telomere lengths. As this happens, the anti-oxidant functions of the body may slowly become less and less capable of dealing with the oxidative damage that occurs. This process then may gradually increase in intensity after young adulthood until it reaches a point where the accumulated oxidative damage deteriorates the homeostatic function of the cells and leads to their eventual death.

Deterioration of the cells and the tissues and organs which are made up of these cells could occur at several different levels. It can occur at the level of the cell membrane which surrounds the cells, or at the level of the cytoplasm which contains the living functional components of the cells, or in the nucleus of the cell (the area of the cell that controls the life functions of the cell and the cell's ability to divide into new daughter cells).

Cellular Apoptosis

It is at the level of the cell nucleus that a special process of regulating the cells has recently been discovered. Cells are now known to contain certain mechanisms that will eliminate that cell if it becomes so deteriorated that it is not safely functional within the body. This process of elimination of cells is known as apoptosis. Apoptosis is a pre-programmed cell suicide, which can occur if the cell senses that its nucleus has become so dysfunctional that it is no longer viable or safe to remain within the body. If this occurs and the cell is able to sense that it is no longer a safe and functional constituent within the body, it will turn on the apoptosis program for self-destruction. This program causes the cell to take itself apart and destroy itself. Once it has been destroyed, cleanup cells within the body search it out and eliminate it. This process of cell suicide helps to protect the body from dangerously deteriorating to such an extent that it is unable to function normally. Without this process the tissues and organs of the body would cease to function. Breakdown of the protective mechanisms of apoptosis is surely a significant part of what happens in the aging process.

Another critical reason for the apoptotic program to be in place within the body is that it may help protect it from the development of cancerous cells. It now appears that the program of cell suicide known as apoptosis, when functional, protects the body from the development of cancer cells. When the body cells detect that they have become defective, the apoptotic process "turns on" and eliminates the cell. So, in other words, if the apoptosis program detects that the cells have developed certain of the propensities it recognizes as cancerous, it will destroy the cell and thereby eliminate the new cancer.

It is interesting to note that of the many differences found to exist between normal cells and cancer cells, one is that in cancer cells the apoptotic program has been turned off or is defective. Perhaps as cells within the body randomly and at a slow rate become cancerous, the majority (or perhaps all of them in normal healthy people) are eliminated through apoptosis. However, in a certain tiny minority of cells, the apoptotic process may be defective. Therefore, in the cells that become malignant the apoptotic process is not able to destroy them. This may then be one of the primary steps in the development of malignant, cancerous tumors. We know that cancer in the body stems from one single malignant cell that developed and was not eliminated at an early stage. This one cell then gradually develops into a cancerous growth, known as a tumor. The lack of an active and viable apoptotic program of cell suicide may therefore be one of the critical elements that allow cancer to develop within the body.

Cancer cells therefore have two unique properties that make them very difficult to kill. They no longer have the apoptotic control mechanism that would cause them to kill themselves. They also contain the enzyme telomerase, which preserves their chromosomes from gradual deterioration as they divide over and over again. This combination of protection from the telomere time clock and a loss of the normal inherent ability to kill themselves when they are found to be defective are two important reasons cancer calls are so deadly and so difficult to eliminate from the body through medical treatment.

Loss of homeostatic controls (control over the basic life functions) in both the cells and in the tissues and organs they comprise, are critical parts of the entire aging and cancer development processes. This is because organs and tissues of the body have controlling and stimulating effects on other organs and tissues. Most particularly, certain areas of the brain produce hormones and other substances that help to stimulate and maintain myriad other tissues of the body. It may be that as the cells of the hypothalamus and pituitary glands of the brain deteriorate they are eliminated through the process of apoptosis. This loss of functioning cells within the control areas of the brain may lead to dysfunction of other organs which they control such as the thyroid, testicles, ovaries and other parts of the brain and hormonal systems which in turn, control the body as a whole. It is possible that a telomere time clock within these tiny areas of the brain actually is the single most critical element that sets the human life span. Perhaps the body ages substantially as a result of the gradual deterioration and loss of cells of the brain. As the cells of the brain's pituitary and hypothalamus tick down their telomere time clock and cease their function after being eliminated through apoptosis, the various stages of adult life cease to be adequately controlled.

For example, by late adulthood, many hormones have dropped nearly to zero, or have been drastically reduced from their original young adult functional levels. These include melatonoin, DHEA, testosterone, estrogen and thyroid hormone. It appears plausible that the gradual loss of these hormones allows the rest of the tissues of the body to gradually deteriorate and go through the process of aging known as senescence.

Death Hormone

Another interesting concept in aging is whether or not a "death hormone" exists. It is possible that somewhere within the pituitary, hypothalamus, and pineal glands, there exists a hormone, which at some point in life is produced, spreads throughout the body, and then causes death. This potential "death hormone" has not yet been discovered but many scientists are avidly searching for it, as they believe that there is significant evidence that it exists. Such a "death hormone," or a series of hormones which act as "death hormones," may actually work on a widespread basis in a way analogous to the apoptosis. The difference being that apoptosis works more locally since it affects only one cell at a time.

Apoptosis, as we mentioned before, is a program of cellular self-suicide, which occurs when the cell reaches a stage of great dysfunction and "kills itself," thus eliminating itself from the body. In a similar manner, the brain may be able to detect when the body has reached such a degree of dysfunction from aging that it should no longer exist. When this occurs, a death hormone, which actually causes the entire body to die, may be released. This theory may be described as an "organismal apoptosis" or a pre-programmed suicide of the entire organism.

Another very significant factor, which may be influenced through life extension, is the augmentation of the hormones that control the organs and tissues of the human body. If humans are programmed to only develop into healthy, reproducing young adults due to the lack of a further hormonal control program, then perhaps we can institute a new program through hormone replacement which will tell the body to continue in a youthful and vigorous condition beyond young adulthood.

One plausible way to do this would be to attempt to replace the known major hormones of the body to levels that are found in young adulthood or early middle age. It may well be that replacing and maintaining these hormone levels will signal the body to maintain itself in that condition of either a young adult or a middle-aged individual. At the present time, life extension via such manipulations of our hormonal control mechanism seems to be the most useful tool available in anti-aging medicine.

Multifactorial Theory of Aging and Cancer

An integrated theory of aging and the development of aging-related degenerative disease can be deduced if we consider all the known mechanisms of cell deterioration that are thought to cause aging. A multi-factorial aging theory based on a combination of random damage, pre-programmed obsolescence, and the lack of a predetermined program for continued development and maintenance of the human body may explain how we age, deteriorate, and die.

Within the random damage portion of this multi-action scheme, free radical damage may be the most significant factor. Damage from oxidative free radicals and other oxidative sources such as radiation, x-ray and chemicals may deteriorate the cells of the body. Along with this damage, a buildup of intra-cellular waste and deterioration of the homeostatic control mechanisms within the cytoplasm of the cell and within the nucleus of the cell, may lead to it becoming terminally dysfunctional. Additionally, a pre-programmed obsolescence in which the telomere time clock gradually deteriorates the cell's ability to function, to maintain itself, to cleanse itself, and to protect itself from free radical damage may gradually develop. This pre-programmed obsolescence may culminate in the telomere ticking down to zero, at which point the cell is no longer able to divide and will therefore be grossly dysfunctional. Any of these significant dysfunctions may then lead the apoptotic processes of the body to kill the cell and eliminate it from the body. The loss of individual cells from certain critical glands and areas of the body may compound the entire aging process by causing the organs as a whole to become dysfunctional.

In particular, dysfunction or apoptotic elimination of cells within the hormonal control areas of our brain would have a devastating effect on the ability of the body as a whole to maintain its normal physiologic functioning. When all of this has reached a climax, it may be that a death hormone is released in response to the brain's sensing of the gross dysfunction. Such a death hormone may then cause the entire organism to die in a similar manner to that in which apoptosis senses dysfunction within a single cell and causes that single cell to die and be eliminated.

Why is it that the body may have sensing mechanisms that would cause the organism as a whole to be eliminated from life? In the same way that apoptosis may sense that significant dysfunction within an individual cell may lead to problems for the organism as a whole if that cell is not eliminated, it may be that the body inherits a program which eliminates it from the reproductive cycle of human life once it has deteriorated to such an extent that it would not be beneficial for the human race to have it continue within the reproductive pool.

Thoughts of a death hormone and rationale for the existence of a death hormone are most certainly speculative, but are very interesting when viewed from the perspective of evolutionary and cell biology science. If the body has been designed, or has evolved, simply to reproduce itself through sexual reproduction, then there would be no need for it to contain any program, plan or mechanisms to preserve itself after children are born and raised to an age at which they can care for themselves.

If human beings become reproductively able in late adolescence between the ages of fourteen and sixteen and if children could reasonably be expected to care for themselves after the age of eight, we would find that humans may not ever have been designed or have evolved any mechanism to ensure survival beyond the age of twenty-five or so. And this, interestingly enough, is exactly the age range at which humans generally begin to show the first signs of aging. Therefore, it may be that there are no design features that tell the body how to preserve itself or how to function after the age of roughly twenty-five. If this is the case, perhaps the decades of life that we live after the age of twenty-five are simply icing on the cake and are an accidental result of the great efficiency that the body has inherent within its design to reach young adulthood. The protective mechanisms that the body has been endowed with to help it remain in a very healthy state in the teens and early twenties may, therefore, ameliorate aging and may be the basis for our slow and gradual decline into aging.

Compared to other mammals and many other animals, human beings have a very long life span past this reproductive time of life. It is interesting to note that humans have the greatest abundance and efficiency of anti-oxidant enzymes of any animal. It may be that these antioxidant enzymes aid the body to be very vigorous and strong, and, in particular, may help preserve brain tissues into young adulthood. Past that point, an incidental result of the strength and vigor of these antioxidant enzymes may allow human beings to live far longer than do mammals that live to reproductive age and then die.

Degenerative diseases seem to be closely associated with the aging process. The incidence of cancer, arthritis, heart disease, blood vessel disease, dementia, and autoimmune disorders is known to increase dramatically in older persons. It seems likely that the mechanisms of deterioration, which cause the aging process, also gives rise to these degenerative diseases. If the deterioration does not directly cause the diseases, then they may allow for their development by weakening the body to such an extent that it is more vulnerable to the development of disease. A review of the scientific literature leads us to the conclusion that many of today's most troubling degenerative diseases are either directly or indirectly caused by the aging process.

In exploring the potential mechanisms for aging, it becomes evident that several different therapies for life extension are plausible. Looking at the random damage portion of the aging process, it seems that augmenting the human body's ability to fight free radical damage may help to decrease its gradual deterioration from this process. The buildup of inter-cellular waste may be diminished through internal cleansing, detoxification, and other therapies, which help rid the body of waste products, and of toxic materials.

The pre-programmed obsolesce portion of the multi-factorial process which is known as aging, may be based on the telomere time clock. Perhaps the most potent treatment for the aging process may therefore occur in the future when a way is discovered to influence this telomere time clock, or influence other related enzymes which also preserve the genes. Several large biomedical companies are presently experimenting with the telomerase enzyme for ways to re-institute telomerase activity within the normal cells of the body and to eliminate telomerase activity from cancerous cells. In the future, if researchers are able to re-institute telomerase activity within the normal somatic cells of the body, the most significant root cause of the entire aging process may have been negated and may eventually be completely arrested.

Homeopathic Telomerase

The newest and in some ways most controversial of the many potential interventions upon the aging process is the use of homeopathic telomerase. Conventional telomerase therapy is quite different from homeopathic telomerase therapy. The conventional therapy will employ viruses to spread the telomerase gene to all the somatic cells of the body. The somatic cells will then become immortalized and very resistant to aging. Such "viral transfection" techniques are currently being developed by the major bio-technology companies. The biotech companies are also experimenting with ways to curtail the telomerase gene in cancer cells. This will cause the cancer cells to become mortal, in that they will have a limited lifespan thanks to the restored activity of the telomere timeclock.

Homeopathic telomerase therapy rests upon a completely different mechanism of action. It utilizes the "vital force" concept to energize the tissues and organs of the body with the power of telomerase. Therefore, in cancerous tissues the vital force will be stimulated to resist the immortalizing effect of the telomerase gene present within the cancer cells. To promote longevity in a healthy individual, homeopathic telomerase may be used to stimulate immortalizing properties within the somatic cells of the body that inherently lack the telomerase enzyme.

Is it plausible to utilize potentized "homeopathic" telomerase as a direct replacement for conventional telomerase therapy and expect to attain the same results? I do not think so. The direct action of placing telomerase genes within the nucleus of cells deficient in the enzyme, and of removing the gene or blocking the gene's activity in cancer cells, will likely have a more profound effect upon the individual. Homeopathic telomerase therapy has a less direct and more subtle effect since it relies upon producing improvements in the body's energy system (also known as the vital force) rather than producing a direct biochemical change in each cell.

Another important concept to consider is how homeopathic telomerase can be produced. If we simply start with purified telomerase, available from pharmaceutical companies, and then potentize it, are we actually producing a homeopathic medicine that may rationally be expected to curtail the aging process? I do not believe so.

My reasoning is that if we consider the fundamental tenet of homeopathy, "like treats like," it makes no sense that potentizing telomerase will make up for its lack in the aging cells of the body. Let me use an example from classical homeopathic medical practice. One of the oldest homeopathic remedies is thyroid. We use it in two forms. The "crude" form is simply whole thyroid gland and is used to replace missing thyroid hormone in those with low thyroid functioning (this is called hypothyroidism and is analogous to the telomerase deficiency found in our aging-prone somatic cells). We therefore administer whole thyroid gland, containing pharmacologically active thyroid hormone when we wish to treat hypothyroidism. We prescribe the other form, called thyroidinum (homeopathically diluted and potentized thyroid gland substance) when we are treating the constellation of symptoms that matches this remedy. We licensed Homeopathic Medical Doctors do not utilize the diluted and potentized form of thyroid (thyroidinum) to treat gross deficiencies of thyroid hormone (hypothyroidism). Instead, we have the patient take the "crude" form of glandular thyroid since it contains sufficient quantities of pure thyroid hormone to replace the missing hormone in the hypothyroid patient's bloodstream.

Following the tenets of classical homeopathic theory, I prepare two kinds of homeopathic telomerase based upon their intended uses. I potentize telomerase and the nuclear cellular components of the telomerase-containing cancer cells and produce a homeopathic remedy to influence the vital force to resist the cancer. This preparation is very similar to the old homeopathic remedy called "carcinosin" except the telomerase component and related enzymes have been magnified.

A second type is prepared from the nuclear components of aging cells and is intended to stimulate the vital force to resist the aging processes that derive from a lack of telomerase enzyme activity in the somatic cells. As a licensed homeopathic physician, it makes sense to me that both conventional and homeopathic telomerase therapy can rationally be used simultaneously since they each work via different mechanisms of action.

References

Katri Koli and Jorma Keski-Oja, "Cellular Senescence," Annals of Medicine 24 (1992): 313-18.

Calvin B. Harley, "Telomere Loss; Mitotic Clock or Genetic Time Bomb?" Molecular Research 256 (1991): 271-82, Daniella Monti, Leonarda Troiano, Franco Thopea, Emanuella Grassilla, Andrea Cossarizza, Daniella Barozzi, Maria Clauda Pellini, Maria Gracia Tamassia, Giorgio Palomo, Claudio Franceschi.

"Apoptosis Programmed Cell Death; A Role in the Aging Process?" American Journal of Clinical Nutrition 55 (1992): 1208S-14S.

Gwyn T. Williams, "Programmed Cell Death" Apoptosis and Oncogenesis Cell 65 (June 28, 1991): 1097-98.

Y. Yin, N. A. Tainsky, F. Z. Bischoff, L. C. Strong, and G. M. Wahl. "Royal Type P53 Restores Cell Cycle Control and Inhibits Gene Amplification in Cell with Mutant P 53 Cell", 1992.

E. Yonish-Rouach, D. Grunwald, S. Wilder, A. Kimchi, E. Maw, J. Lawrence, P. Maw, N. Oren, "P53 Mediated Cell Death; Relationship to Cell Cycle Control," Molecular and Cellular Biology (March 1993): 1415-23.

SECTION TWO - DETOXIFICATION

Chapter Three
DETOXIFICATION IN AGING, CANCER AND HEART DISEASE

Detoxification is the process of the removal of poisonous, noxious, and otherwise damaging materials from the body. When toxic materials from the outside world enter the body, they can accumulate to harmful levels. Toxic materials can include chemicals, metals and pesticides that are absorbed or ingested into the body and then deposited into the body's tissues. Other sources of toxic substances include waste byproducts from normal body metabolism, and abnormal deposits of materials (such as copper, iron and calcium) which otherwise would be beneficial to the body. In a manner similar to the inappropriate deposits of otherwise normal metals, the body may also choose to deposit materials such as the protein material known as "amyloid" in abnormal quantities and in unusual and improper places. Removal of amyloid, toxic metals, abnormal deposits of normal metals and toxic chemicals can be beneficial to health and longevity.

Specific organs seem to be particularly prone to damage and to diminished viability through the process of accumulation of toxic materials. These particularly sensitive organs include the liver, gall bladder, kidney, brain, heart and vascular system. As toxic materials collect in these organs, the organs' functional capability is diminished.

The liver may be prone to accumulate toxins due to the fact that one of its roles in the body is to screen out and attempt to detoxify such materials. Therefore, the liver will tend to trap and collect toxic materials. Its enzymes will be put to work in an effort to detoxify those materials. However, when the liver reaches a certain toxic burden, it is unable to completely detoxify and remove all these materials and may, therefore, slowly build up a burden of toxic materials.

The gall bladder is connected to the liver. It stores and excretes the bile produced by the liver and may also be burdened by toxicity. The kidneys, which are the other major filtering organ of the body, can also slowly collect the toxic materials, which it is trying to eliminate through the urine. The tissues of the brain, heart, and vascular system tend to collect toxic materials and their functions are diminished by the deposit and retention of toxic materials.

The genetic aging of each cell may potentially be linked to harm caused by the gradual accumulation of intra-cellular toxins over time. There is a distinct possibility that as our cells age genetically, they are less able to rid themselves of toxins. Perhaps the toxins themselves increase the rate of genetic aging of each cell. Telomerase therapy offers an interesting adjunct to detoxification regimens. Telomerase may allow the each cell to retain its youthful resistance to the accumulation and harmful effects of toxins

Overview of Metal Toxicity

Metals represent a significant category of potentially toxic materials that may require removal through the detoxification process. Metals requiring detoxification can be placed into three general categories. The first of these contains toxic metals such as lead, cadmium, and mercury. The second category includes toxic levels of otherwise beneficial metals. Examples of these are iron and copper. Thirdly is metastatic calcium. Calcium is technically known to chemists as a metal. It is termed "metastatic" if deposited in unhealthy locations. Therefore, we have the three general categories to discuss: 1) Metals which in and of themselves are toxic, 2) metals which are needed by the body and are only toxic when deposited in toxic levels in the body, and 3) metastatic calcium which is harmful and toxic to the body when it is being deposited in large enough amounts and in improper locations so that it interferes with the normal function of the tissue in which it has been deposited.

First under discussion are the metals which, in and of themselves, are toxic and have no known benefits to the body's normal functioning. High levels of lead have been found in American children and throughout the population in all age groups. Lead is a metal that tends to accumulate in the bones, the kidneys, and in the brains of people exposed. Unfortunately, in modern industrial society, many are so exposed. Perhaps only a minority has such a significantly toxic burden of lead that the body becomes openly symptomatic with disease. However, lower amounts of toxic lead deposits are likely to interfere with the normal physiological functions of the body and thereby worsen the normal aging process and make more likely, or more severe, the common degenerative diseases of aging. Lead is well known to be a co-factor in the development and progression of high blood pressure. It is also likely to play a role in dementia and other deterioration of the brain and nervous system. Lead may also play a role in heart disease. The toxic metal cadmium is also commonly found in those suffering from lead toxicity. Cadmium is another toxic heavy metal that seems rather widespread in the environment due to industrial processes and the burning of leaded gasoline. Lead and cadmium act together in similar ways and are know co-factors in high blood pressure. It seems very likely that they work together to increase the rate and severity of the generalized aging process.

The toxic metal "mercury" has been known to be very dangerous for many decades. During the last century when the metal was used without precaution in industrial processes, many industrial workers became severely ill due to their exposure to the noxious metal. One of the most famous examples of this occurred in the hat manufacturing industry. The term "mad as a hatter" comes from the fact that so many hat makers were afflicted with mental disease secondary to their chronic exposure to high levels of mercury. The mercury was taken into the body through skin contact and likely also through inhalation of mercury vapors. As the workers slowly accumulated mercury within their brains, its poisonous effects caused them to go insane or to become demented.

In modern society, mercury is used in many industrial processes and is present in many products, which we come in contact with daily. Since mercury is a known toxic metal, it is usually placed in products in such a way as to protect the user from the mercury itself. However, this is not always one hundred percent successful. Americans may come into contact with mercury more often than they realize.

Perhaps the leading source of mercury toxicity in humans comes from the use of mercury-silver amalgam fillings in teeth and other dental work. Amalgam fillings have been the most common type of dental fillings in the United States and throughout the western world for most of this century. Amalgam fillings are made of an alloy of several metals, the most prominent being mercury and silver. Approximately fifty percent of the material that comprises metal amalgam fillings is mercury. Mercury, along with copper, silver and other metals form an alloy that is very easily used by the dentist. It is easy to produce the alloy and it is easy to form it into the areas that have been drilled out for fillings. The material has also been found to be fairly long lasting and durable. Mercury-silver amalgam fillings often last nearly a decade, sometimes longer. It has therefore been very popular as a material for dentists to use. However, the fact that it contains such a significant percentage of mercury raises health concerns.

Dentists have long held that the mercury in amalgam fillings is so tightly bonded to the other metals in the alloy that it does not escape and is not absorbed into the body. However, recent research indicates that there is evidence that the mercury within amalgam fillings does find its way into the body and is deposited in various organs and tissues. The most problematic deposits occur in the brain and nervous tissue.

There has been much speculation about exactly how mercury finds its way from the fillings in our teeth to the other tissues in the body. Perhaps mercury is slowly absorbed through the teeth, then into the saliva, and finally into the body. Another very likely means of mercury toxification into the body is through the constant inhalation of mercury vapors that emanate from mercury fillings in the teeth. It has been shown that chewing creates pressure that is exerted on the fillings. This releases mercury vapor as a result of the pressure and the heat of the chewing action. We then breathe the vapor, which can easily enter the bloodstream. Mercury vapors seem to be released more readily when the filling is newer. As the fillings age, the mercury component is gradually lost and the fillings become brittle and eventually decay. This gradual decay of the amalgam fillings is one of the prime reasons to have them replaced every decade or so.

Mercury is a known contributor to diseases such as high blood pressure, heart disease, dementia and other afflictions of the nervous system. Some researchers believe that mercury toxicity is a causative factor in multiple sclerosis. MS is a degenerative nerve disease. Toxic burdens of mercury likely tax the ability of the cells of the kidney and the liver to detoxify other materials with which they come in contact.

Toxic levels of otherwise beneficial metals such as iron and copper can also occur. Iron and copper are essential metals the body requires for its normal physiological functioning. There are special absorption mechanisms within the intestines for bringing copper and iron into our body. Likewise, special protein transport molecules are present in the blood to transport iron, copper and other necessary metals to tissues of the body where they are needed.

Iron and copper are certainly essential metals to maintain a healthy body. Iron is used primarily as the centerpiece of the hemoglobin molecule. Hemoglobin molecules are the oxygen carrying portions of our red blood cells. Without iron, hemoglobin would not be made. People who are iron deficient can develop anemia. Anemia is a lack of hemoglobin and red blood cells resulting in a decreased ability of the blood to carry and transport oxygen to the tissues for metabolism. Even though iron and copper are necessary for the body's health and well being, deposits of abnormally high amounts of iron and copper can actually poison the body and lead to disease.

Wilson's disease is caused by abnormal deposits of high amounts of copper. Hemosiderosis is one of the diseases caused by abnormal deposits of high amounts of iron. It is very likely that abnormally high deposits of iron and copper increase the rate of the aging process. It is postulated that this occurs due to increased production of free radicals in the body. Free radicals are those reactive chemical species that have a propensity to interact with the healthy tissues of the body and deteriorate them through the process of oxidation. Oxidation is the process that occurs during rusting. When metal is exposed to rain and air, it will deteriorate by forming rust. This rust is the byproduct of the oxidation of the metal by the oxygen in the air. The presence of abnormal deposits of iron and copper directly leads to an excess production of free radicals. They then circulate and damage the tissues of the body. Therefore, iron and copper levels should be maintained at the proper physiological levels.

Unusual deposits of calcium within blood vessels, ligaments, and tendons can lead to disease. They can accelerate the aging process within these tissues and within the body in general. Such unusual and inappropriately deposited calcium has been known, for quite some time, as a source of aging-related diseases. The famed scientist Hans Selye's book, *Calciphylaxis,* focused on abnormal deposits of calcium and its detrimental effects on the body's health. His premise was that, over time, calcium is slowly deposited in areas of the body where it causes harm and detriment. At the same time, calcium is lost from the bones and other tissues of the body where it is essential as a building block and needed to maintain good health. This abnormal distribution of calcium, in his opinion, leads to many of the degenerative diseases that occur along with aging. He also surmised that abnormal deposits of calcium might be a cause of the aging process itself.

Calcified deposits are most infamous for their effects on arteries. Arteries often afflicted by abnormal calcium deposits include the arteries of the heart, the arteries of the neck that allow blood flow to the brain, and the very large arteries of the central body which include the aorta and the femoral arteries of the legs. Substantial calcium deposits within the arteries are usually associated with fatty plaques that surround the calcium. The entire complex of abnormal calcium and fatty deposits is known as an atheromatous plaque. Atheromatous plaques are the main constituents of the blood vessel disease known as atherosclerosis. Atherosclerotic plaques build up within the walls of these essential, large arteries and cause them to be blocked off. When the arteries reach a state of being blocked to a significant extent, symptoms and diseases such as angina, heart attack, strokes and lack of blood supply to the muscles of the leg known as "intermittent claudication" can occur.

Harmful plaques develop over time. As they develop over the course of decades prior to becoming symptomatic, calcium, along with fatty material, is slowly deposited. The calcium tends to make the plaques stiff and hard. This is particularly true of the areas where the calcium is deposited diffusely in the artery walls of much smaller arteries. The widespread calcium deposits and the resultant disease of the blood supply involving the small arteries is known as "arteriosclerosis." Arteriosclerosis is commonly called "hardening of the arteries." This is certainly an apt term since that is exactly what happens. Arteries are rubbery tubes through which blood flows. These rubbery tubes have areas of muscle within their rubbery walls. The muscle helps the arteries either contract to a smaller diameter or expand to a larger diameter. This expansion and contraction of arteries occurs when different physiological states require either more, or less, blood flow.

When we eat a meal, for example, the small arteries that supply blood to the muscles of the arms and legs constrict so that they do not allow as much blood to flow to these large muscles which are not being used. The blood is thereby shunted to the arteries that supply the digestive organs of the body. Several hours later, after the food has digested, the exact opposite happens if the individual begins to exercise. The small arteries that supply blood to the digestive organs constrict down so that they limit blood flow to that area while the small arteries supplying blood to the muscles of the arms and legs dilate, allowing much greater blood flow to the muscles during exercise.

Hardening of the arteries causes these rubbery tube arteries to become so stiff that they are very nearly non-pliable. This non-pliability makes them stiff, so that their muscular walls are not able to constrict and dilate as needed. This, of course, limits the oxygen and nutrients contained within the blood flow to all the various areas of the body which are afflicted with the arteriosclerosis. The stiffness of the small arteries may also be important as a cause of high blood pressure (hypertension).

The limitation of blood supply that arises both from atheromatous plaques and arteriosclerosis can cause the organs of the body to lack the proper oxygen and nutrients for metabolism. This may increase the ability of free radicals to cause damage to the tissues of the afflicted areas. It seems plausible that arteriosclerosis and atherosclerosis, both of which have metastatic calcium as an integral feature, pose a threat to both health and longevity.

Treatment of Metal Toxicity

Detoxification of heavy metal substances and calcium is an important concern in anti-aging medicine. Heavy metals that may build up to toxic levels in the human body include arsenic, cadmium, lead, mercury and thallium. These heavy metals, when present in the human body, may act to decrease the normal physiological processes of the body that help to maintain longevity. Heavy metal toxins can directly poison or inactivate critical enzyme systems within the cells. Heavy metals may also disrupt functioning of tissues and even whole organs of the body. The functions of the brain and nervous tissues may be impaired, the filtering ability of the kidneys may be damaged, and the heart muscle may be poisoned and therefore weakened. Either directly or indirectly, several of these heavy metals have been associated with hypertension. Hypertension is one of the most common problems afflicting modern man.

Lead represents perhaps the most common heavy metal that accumulates to toxic levels in the human body. The lead that collects may come from natural and man-made sources. Small amounts of lead are found in the soil and rocks of the earth and in small amounts throughout nature. However, the increased use of lead as an industrial metal has radically increased the lead exposure of human beings. Lead from leaded gasoline can collect in soils and be ingested by children. Likewise, children can ingest lead through the eating of paint chips that contain lead. Therefore, plumbism, which is the name for childhood lead poisoning, is prevalent in industrial society. Certainly, much of this plumbism goes untreated and adults may therefore be carrying a toxic burden of lead. It is estimated that as many as two million children per year are afflicted with plumbism in the United States.

Salts of lead are absorbed through either ingesting them or inhaling them. Similarly, certain organic forms of lead salts can be directly absorbed through the skin. The majority of ingested lead is taken up by the bone and integrated into the crystalline structure of the bone. Five to ten percent of ingested lead is maintained in the blood stream. Smaller percentages of ingested lead are deposited in the soft tissues of the body, including the brain and kidneys. The body tries to excrete the lead burden in the stool, urine, hair, nails, sweat and saliva.

Lead poisoning causes damage in at least two ways. It poisons enzymes by binding to the sulfhydryl group of proteins. Lead can also, when in high concentrations, alter the ability of proteins that make up the cells to mold into their correct shape. This acts to denature the protein and causes both death of cells and inflammation of the tissues that those cells comprise.

Symptoms of acute adult intoxifications from high amounts of lead include abdominal pain, headache, memory loss and changes in the ability of the peripheral nerves to function. Clinical signs of acute high dose lead poisoning in adults include renal disease, anemia, and "demyelination" or destruction of the coating of certain nerves.

Perhaps more commonly, adults are afflicted with a subclinical form of lead poisoning. In a sense we mean that they are negatively affected by the lead poisoning but not to such a degree that sudden or acute signs and symptoms occur. Certainly, one of the most common subclinical forms of chronic lead poisoning in adults is high blood pressure. Other findings of chronic lead poisoning in adults include decreased kidney function, decreases in the various functions of the brain, and deterioration of the nervous system in the extremities.

Laboratory testing for chronic lead poisoning includes hair analysis. Hair analysis involves removal of a small sample of hair from the scalp. The sample is then analyzed for the small amounts of lead that may accumulate within the hair. Lead can also be detected by tests of the urine. Urine tests for lead can be performed either on routine urine samples or on urine that is collected after medicines that cause the body to increase its elimination of lead have been administered.

Mercury is encountered often by modern humans. There are three common toxic forms of mercury to which we become exposed. They are inorganic elemental mercury, inorganic salts of mercury, and organic mercury. Organic mercury is most commonly encountered as the methyl form of mercury. Of the three toxic forms of mercury, organic mercury is the most widespread and probably the most dangerous form. Mercury in its elemental form is found in dental amalgams and other man-made materials. Salts of mercury are also found in some medicines, particularly those that are meant to be applied to the skin, and in some foodstuffs. Paints, fungicides, treated seeds and cosmetic agents are also known sources of organic mercury.

The elemental form of mercury is poorly absorbed through the intestinal tract but can be absorbed easily as vapor in the air. Inorganic salts of mercury are more readily absorbed through the intestinal tract and also through the skin. Organic methyl mercury is absorbed very easily through the intestines and the skin. Once mercury is absorbed into the body it can dissolve very easily into fatty or "lipid" components of the tissues. Mercury can easily cross the blood brain barrier and enter the brain cells. Once absorbed into brain cells and other cells of the nervous system, mercury remains strongly adherent to those tissues. Brain and nervous system cells are perhaps the most difficult tissues from which to remove mercury.

Another place where mercury deposits is in the kidneys. Just as does lead, mercury may poison and interfere with the normal functioning of the kidneys. The kidneys hold a key role in the general health of the body by being one of its primary organs for the elimination of waste products. Chronic poisoning from inorganic forms of mercury can result in neurologic symptoms from mercury's effects on the central nervous system (brain). These symptoms can include tiredness, weight loss, loss of appetite and difficulty with digestion. Similarly, memory loss, insomnia, hyper excitability and, in some cases, excessive timidity, can result from chronic exposure. Organic forms of mercury cause similar symptoms.

Treatment of chronic mercury poisoning can be achieved through eliminating sources of mercury absorption and by enhancing the elimination of mercury. One of the most common sources of chronic mercury exposure treated by alternative medical practitioners is that of mercury silver amalgams in the teeth. Perhaps the most common type of dental filling for cavities and other uses, mercury silver amalgams contain roughly fifty percent mercury by weight.

Dentists maintain that this mercury is bound to the silver and other constituents of the amalgam filling and is unavailable to be absorbed into the rest of the body. However, much controversy surrounds the question of the possible absorption of mercury into the rest of the body. Because of this, some physicians now espouse the removal of mercury silver amalgam fillings and are replacing them with nontoxic substances. Chelation therapies can be used to rid the body of excess mercury accumulations. Penicillamine, dimercaprol, DMPS, and DMSA, are chelation medicines for mercury toxicity.

Cadmium is another fairly common source of heavy metal poisoning. Cadmium can be found in some shellfish, in pollution secondary to mining and smelting operations, and in batteries, ceramics, paints, plastics, and electroplating. Roughly half of the cadmium which is absorbed is concentrated in the liver and kidneys. Chronic cadmium intoxication usually results from inhalation of cadmium from industrial sources.

Chronic intoxication can cause emphysema of the lungs and kidney damage. Hypertension is another associated finding of chronic intoxication by cadmium. Urinary tests for cadmium can reveal its presence. Likewise, hair analysis can reveal hidden chronic cadmium poisoning. Chelation agents are known to bind cadmium. EDTA and DMSA are the chelating agents most commonly used in the treatment of both chronic and acute cadmium intoxication.

Arsenic compounds can occur in insecticides, rodenticides, fungicides, herbicides and wood preservatives. Arsenic can be found as an industrial pollutant, in certain medicinal compounds, and in contaminated food and water. Absorption of arsenic can occur through the lungs, the skin and the gastrointestinal tract. Arsenic can be clinically detected by testing the hair and nails. Ailments resulting from chronic arsenic intoxication include abnormal liver functioning, anemia, kidney disruption and resulting urinary abnormalities. Common chelating medicines are able to remove arsenic by producing a complex of arsenic that is excreted by the kidneys.

Thallium is a toxic metal that has been used as an insecticide, as a rodenticide, in fireworks, and in many kinds of industrial processes. Thallium can be absorbed through the skin, by inhalation and through the gastrointestinal tract. Thallium can cause disease in humans by interfering with a critical biochemical process known as "oxidative phosphorylation." It does this through its inhibition of an enzyme called ATPase.

Humans who have been examined after death from thallium poisoning reveal swelling of the brain, loss of the coverings of the nerves known as "myelin sheaths" in their extremities, fatty deterioration of the liver, and degeneration of the heart muscle. Therapies for acute thallium poisoning include cleansing of the gastrointestinal tract, dialysis, and medicines to enhance the excretion of thallium by the kidneys.

Calcium can be removed from areas in which it has been inappropriately deposited. Chelation therapy may help to remove such abnormal calcium deposits and thereby improve the body's health. Useful chelating agents for calcium include both natural and man-made substances. Nutritional substances such as chlorella, spirulina, and garlic have long been regarded as agents that can help to detoxify the body of abnormal calcium and toxins. EDTA chelation therapy is also used by medical practitioners in an attempt to remove unwanted calcium deposits. Improvements in cardiovascular health following EDTA chelation therapy may result from both the removal of metastatic calcium and other heavy metals.

General Systemic Detoxification

All of the tissues and organs that comprise the human body can become burdened with toxic wastes and poisons. We have previously discussed heavy metal toxins, abnormally deposited calcium, and the build-up of cellular waste as worthy targets for medical detoxification. However, other toxins such as insecticides, herbicides, petroleum derivatives and other environmental poisons can also afflict the body.

It is interesting to note that although we commonly think of man-made substances as the sole source of toxins, nature also confronts us with a substantial array of noxious or poisonous compounds of which the body's detoxification aparati must contend. Molds and fungi on our food grains produce cancer causing toxins such as Aflatoxin. Many species of plants produce mild poisons that are present in the outer layers of the fruit. Potatoes and tomatoes are both members of the nightshade family of plants and each contain mildly poisonous chemicals within their outer skin. Livers and kidneys must have the functional capacities to cope with these toxins. If they themselves become overly burdened with retained toxins, they lose much of this ability to detoxify the blood and thereby protect the rest of the body.

The liver is particularly vulnerable to toxins since it filters the blood and serves as the largest detoxifying organ in the human body. The liver cells are programmed to filter-out and destroy toxins via a complex set of detoxification enzymes. If the liver retains too many toxins, it will itself become poisoned and may lose its ability to detoxify and cleanse the blood. Worse still, retained toxins may completely destroy the liver, as occurs in alcoholic cirrhosis, or may cause cancer to develop within the liver.

Similarly, the kidneys also filter the blood and remove normal waste products as well as toxic substances. Damage to the kidneys by retained toxins can result in consequences similar to those that may occur in the liver. It is important that both of these crucial detoxification organs be maintained in good condition and with the least accumulation of toxins possible.

How can this be achieved? A first line strategy is to attempt to keep as many toxins as possible out of the body. Changing our habits so that we minimize exposure to pollutants, chemicals and the toxins found in foods allows the liver and kidneys a greater chance to cleanse themselves. Next, we can utilize nutritional substances such as garlic, parsley, alfalfa, spirulina, chlorella, and many of the chlorophyll containing plant substances to naturally cleanse the system. Certain medical techniques such as colon cleansing and intravenous therapies may serve to rid the body of damaging toxins and waste products. Telomerase therapy is an interesting new possibility as a detoxification therapy. Perhaps it will rejuvenate and strengthen the ability of each cell to rid itself of toxins, and to protect itself against the harmful effects of toxins.

Detoxification Case History

Mr. J.C., a man in his forties, reported to our LIFExtension Center Reno medical clinic in early 1999. He complained of having major problems with high blood pressure and abnormal heartbeats. He had been under the care of an internist for more than one year, but was very unhappy with side effects from the multiple medications on which the doctor had placed him. I immediately tested him for toxins by ordering an analysis of his head hair. The test results showed him to have high levels of toxic mercury and other toxins that may damage the heart and blood vessels. The test also revealed him to be low in certain minerals essential to the maintenance of normal blood pressure and heart rhythm, such as magnesium, calcium and potassium.

He was then begun on an intensive course of detoxification, while at the same time being slowly weaned off of his medications. Within three weeks of this program of intravenous detoxification treatments, oral detoxification supplements and mineral replacement, he was found to be free of heartbeat abnormalities. His blood pressure had also dropped significantly into the top of the normally acceptable range. This in spite of the fact that he had been taken off of his medications. As of today, April, 2000, he is doing well with intermittent maintenance detox treatments and is not using any medication at all.

How To Detoxify

There are many forms of toxins that can damage the human body and impair our mental functioning. If significant toxicity is suspected, it is important to allow a qualified physician to evaluate the case. However, here are some simple ideas to consider using for preventive detoxification:

1) Eat garlic regularly, or at least on an intermittent basis. Garlic appears to cleanse the arterial system, the blood and many other tissues of the body. It is interesting to note that an important part of the healthful "mediterranean diet" is the daily consumption of garlic.... and you won't need to worry about vampire attacks either.

2) Consider a three-week course of intensive herbal detoxification two or more times per year. I like to use a combination of garlic, parsley and alfalfa, which can be added to foods, or you can make capsules from the components (health food stores carry gelatin capsules which you can fill with the herbs). Normally, I use a ratio of 4:1:1 garlic:parsley:alfalfa and take two to six capsules three or four times each day for three weeks. It is important to drink plenty of water during any detox regimen and to consider whether you may have an allergy to any of the herbs being used. We want to avoid consuming herbs to which we are allergic.

Chapter Four

MERCURY TOXICITY
ITS ROLE IN CANCER AND HYPERTENSION

Toxicity from the metal "mercury" has been well known since the inception of the industrial revolution. Hat makers and dye-stuffs workers were some of the first to suffer from its harmful effects. The pathology produced by significant exposure to mercury was so readily apparent that workplace protections against mercury exposure were instituted long before most other contamination and pollution control measures were promulgated.

Today, it is rare to see such gross human contamination although it can occur through industrial accidents and pollution. However, the entire population of the developed world is now likely being exposed to greater amounts of mercury than were their forebears in the pre-industrial age. How are we affected by exposure to toxic mercury and what therapeutic measures can be taken in response? What is the evidence for mercury as a causative agent in hypertension and cancer? Let's begin by discussing the metal and the known exposures to it.

Mercury occurs as both natural and man-made forms. Natural and man-made versions can occur as the elemental metal or other forms into which it has been converted. Mercury can be found in several different forms within the living environment. Naturally occurring mercury is fairly widespread, being distributed in soils, rocks, air, and water. A naturally occurring mercury cycles through the biosphere as mercury containing rocks eroded by water, and as gaseous mercury vapors are emitted from volcanic vents. Naturally occurring mercury is usually found in very minute quantities. Man-made mercury pollution is concentrated primarily in areas where it was originally deposited, and in life forms highest in the food chain.

Mercury can be found in both organic and inorganic forms. When combined with other naturally occurring chemicals such as carbon, oxygen or chlorine, mercury can become organic mercury through its incorporation into biological tissues. Methylmercury is perhaps the most common type of toxic organic mercury. Inorganic or metallic mercury is also fairly widespread and may be found as a single chemical element or in combination with the aforementioned elements of carbon, oxygen or chlorine. Organic mercury in the form of methylmercury can accumulate in animals, particularly fish. Fish seem to be able to concentrate organic mercury whether it was from man-made pollution or from natural sources.

Mercury can change from inorganic to organic forms. Portions of the inorganic mercury, released as industrial pollutants, will be converted through microorganisms and other biological processes into organic mercury such as methylmercury. Methylmercury is perhaps the most toxic to human beings because it is most easily assimilated by the body. Pure metallic mercury is less readily digested and absorbed into the human tissues via an oral route, but may be absorbed easier if inhaled as a vapor or if the metal directly contacts skin.

The Environmental Protection Agency has published studies and issued reports that quantify mercury contamination and our exposure to it (EPA, 1971a; EPA, 1971b; EPA, 1975; EPA, 1984; EPA, 1987). Our air has been found to contain approximately 2.4 parts of mercury per trillion parts of air. Much higher levels are found near some industrial areas. The air concentration near mines and refineries that emit mercury has been measured as high as 1800 parts per trillion. Mercury found in the air is usually found in the inorganic form. Levels of mercury measured in water have ranged in levels as great as 500 parts per trillion to 25 parts per trillion. The upper levels, 500 parts per trillion, have been found in some drinking water wells near heavily polluted areas. Average measurements of mercury near "Superfund" and other industrial clean-up sites have averaged 200 parts per trillion in ground water. Normal levels of mercury from water in areas that are thought to be non-polluted measure 25 parts per trillion.

Human exposure to toxic mercury can occur in several different ways. Induction of mercury into human tissues can occur through breathing polluted air, direct dermal contact, dental materials, and through ingestion of contaminated drinking water and foods. By far the most significant way that human tissues become contaminated with mercury is through the ingestion of organic mercury compounds. The consumption of large amounts of fish can cause a significant exposure to organic mercury. Methylmercury, the most common kind of toxic organic mercury, occurs most prevalently in certain species of fish, such as tuna and swordfish.

Water contaminated with inorganic mercury has also been found to be a cause of mercury toxicity. Drinking water contaminated with inorganic mercury salts has been detected in wells adjacent to national priority lists toxic waste sites. Ingestion of this contaminated water has been a source of mercury toxicity of those living in the affected areas. The inhalation of organic mercury vapors can occur by accident if items containing liquid mercury or mercuric gases are accidentally broken or disrupted. Thermometers, as well as many other household and industrial appliances, contain inorganic metallic mercury.

The workplace environment can also allow exposure to toxic mercury. Chemical workers, metal processing personnel, automotive workers, those in the building trades, people involved with the manufacture and installation of electrical equipment, dental care workers, medical care workers, and other health care personnel are the most commonly afflicted by workplace exposure to toxic mercury. Likewise, those who live or work near areas in which environmental pollution with toxic mercury has occurred, such as water and air found near toxic waste sites and chemical spill areas, have also been found to become contaminated with mercuric compounds both organic and inorganic.

Another potential source of human contamination with mercury is from dental materials (Enwonwu, 1987). Minute quantities of metallic inorganic mercury vapor are slowly released from silver mercury dental amalgams. This occurs particularly after the first year or two after new mercury amalgams are placed in the mouth.

The "outgassing" of mercury vapors occurs primarily during chewing when the pressure and heat of the chewing process causes a breakdown and disruption of the amalgam materials. In this manner, small amounts of mercury vapor are released and breathed in. In a similar fashion, tiny amounts of mercury released from deteriorating silver mercury dental amalgams can combine with oral bacteria to form organic mercury compounds. As the metallic mercury in dental amalgams is slowly converted by bacteria in the mouth into organic mercury, the body is constantly exposed to this more toxic form of the metal. Each of the inorganic outgassed mercury vapors and organic mercury compounds can enter the body and be distributed through various tissues and organs.

Both long-term and short-term exposure to the various forms of mercury can cause damage to the human body. Compilations of data regarding such toxic effects can serve as reference (Weiss, 1986). Organic and inorganic mercury toxicity can affect the brain, developing fetus, and kidneys. Researchers have found that organic mercury consumed in contaminated fish and grain causes greater damage to brain and nervous tissues, and to developing human fetuses.

These exposures to organic mercury in contaminated fish and grains have been found to have a lesser effect on the kidneys of those individuals who have been contaminated. Conversely, it has been found that inorganic salts of mercury ingested via contaminated water or food causes greater harm to human kidneys. Long-term maternal exposure to organic mercury compounds leads to brain damage in offspring. Long-term exposure of metallic mercury vapor in adults leads to tremors, shakiness, kidney disease and memory loss. Human exposure to high levels of inorganic and organic mercury on a short-term basis has similar effects. The natural elimination of mercury from the body can occur through the urine and feces. It has been shown that short-term mercury exposure is more readily eliminated from the body than is the toxicity of long-term exposures.

Epidemiological animal studies have allowed for the determination of toxic mercury exposure. Levels in which exposure to either organic or inorganic mercury cause increased levels of health risk have resulted from these studies. Minimal risk levels are those levels of exposure that have been determined to cause detrimental effects. The government's published minimal risk levels for mercury toxicity were derived from studies in which animal species particularly sensitive to mercury toxicity were exposed to the most toxic forms of mercury.

Minimal risk level analysis takes into account known non-cancer toxicity to mercury. However, they do not take into account any potential risk for developing cancer from exposure to mercury. The government uses minimal risk levels as a guideline for human toxicity from the mercury. Therefore, the government believes that if humans are exposed to mercury at a level below the minimal risk level, there will be no harmful health effect.

Certain potential problems with the government's use of minimum risk levels as the guideline of human toxicity exist. The first concern with this system is that the minimum risk levels were determined from non-human animal studies. Although it may be unrealistic to test humans with known toxic materials such as mercury, any extrapolations from animal models cannot, with certainty, indicate toxicity levels for humans. One example of this would be in brain toxicities due to mercury. Human cognitive functions are not exactly duplicated in animal species that are studied. Therefore, it may well be that humans will exhibit noticeable cognitive disruptions at doses different from the minimum risk levels determined by observing animal behavior.

Cognitive functions including short-term and long-term memory, reasoning, thinking, and personality characteristics may be much more difficult to determine in mice than in humans. Another concern is the difficulty in measuring or accurately estimating the true exposure that humans are encountering. In most cases of known mercury poisonings it is not feasible to obtain an accurate estimate of exact levels of mercury contained within the contaminated food or water. Studies that have attempted to determine an accurate estimation of mercury consumption of individuals exposed to organic mercury by the ingestion of contaminated fish and grain have not led to numbers which scientists can agree upon with any degree of certainty.

Let us now look at some specific information from government risk level publications (PHS, 1989). Short-term exposure in humans has been determined to lead to the following effects (as reported mercury concentrations in air and lengths of exposure in hours). Chest pain, shortness of breath and coughing are reported for exposures of three hours to mercury levels in air at 0.13 ppm. Longer term exposure of eight hours at higher levels (5.4 ppm) results in irritability, decreased ambition, and decreased libido. Longer term exposure to mercury toxicity by breathing mercury vapors has resulted in the following data. A fifteen-year length of exposure to a concentration in the air of mercury at 0.0032 ppm has led to persistent shakiness and tremors.

The government estimates the minimum risk level based on humans for long-term exposure of mercury in the air to be 0.000032. When animals have been exposed to mercury on a short-term basis, the detrimental effects have included heart, kidney, lung and stomach damage, liver disease, and possible brain damage. Detrimental health effects of long-term exposure to animals (long-term being defined as longer than fourteen days of exposure) include brain disease, kidney and heart disease, decreased reproductive functioning, tremors and learning disabilities.

Detrimental health effects in humans from breathing organic mercury, as opposed to inorganic metallic mercury, have not yet been accurately determined in government studies. The same is true for animal health effects from breathing organic mercury. Potential detrimental health effects to humans from both long and short-term exposures to inorganic mercury in foods and water have not been accurately determined. Based on extrapolation of animal study data, the government has established a minimum risk level of 0.814 ppm of mercury for short-term exposure (less or equal to 14 days of exposure). Similarly, long-term exposure (more than 14 days exposure) minimal risk levels have been declared to be 0.063 ppm of mercury in food.

Animal data, from which the previous risk levels were determined for the ingestion of inorganic mercury in the consumption of food or water by animals, has yielded more specific data. Short-term exposure data reveals that water levels of mercury of 51.8 ppm over a one-day exposure results in the death of young rats. One hundred sixty-two ppm of mercury in one day results in the death of fetal hamsters. Long-term exposure of animals to inorganic mercury has shown that a two-year exposure at 25.2 ppm leads to kidney disease in rats. Immune system dysfunction in mice was noticed after seven weeks of water concentration of 11 ppm of mercury. Ten-month exposure of mice to mercury concentrations in water of 11.7 ppm led to decreased appetite, behavior abnormalities and weight loss.

Human consumption of organic mercury via food or water is thought by the government to be harmful if it occurs on a short-term basis above their estimated minimum risk level of 0.0027 ppm of mercury. The animal data from which the human minimal risk level for ingestion of organic mercury was derived yields specific information on mercury toxicity in those animals tested.

Short-term exposure of animals to varying levels of organic mercury in water yields further clues as to potential human reactions. At lower doses, problems ranging from behavioral abnormalities in the offspring of exposed rat mothers to male infertility were seen. The death of fetal mice, brain cell death among rabbits, and learning and behavioral problems in adult rats occurred at higher doses. Long-term exposure of animals to ingested organic mercury results in brain, kidney and liver disease in various animals including rats and kittens.

Although it is not ethically or legally possible to test toxic mercury purposefully in humans, studies involving the death of humans after accidental mercury exposure have been published. Acute exposure to high concentrations of vaporous metallic mercury lead to death in humans secondary to respiratory dysfunctions resulting from severe lung tissue damage (Campbell 1948; Matthes et. al. 1958; Teng and Brennan 1959; Tennant et al. 1961). One study reports human demise secondary to an intermediate length of exposure to organic mercury. In this instance diethylmercury was inhaled as a vapor in concentrations ranging from 1 to 1.1 mg per cubic meter of air during a period of four to five months. Autopsy of the subject revealed gastrointestinal disorders (Hill 1943).

Inhalation or exposure to vapors of metallic mercury can have negative cardiovascular effects in humans and in animals. Blood pressure rates have been shown to be higher in humans acutely exposed to mercury vapor. In young children acutely exposed to inhalation of metallic mercury vapor, autopsy revealed dilation of the heart's right ventricle (Campbell 1948; Hallee 1969). When animals were exposed to mercury vapor inhalation, necrosis or deterioration of the heart tissue was seen. Male rats exposed to daily oral doses or organic methylmercuric chloride over a period of nearly a month were found to have increased systolic blood pressure. This increase in blood pressure began approximately two months after the initiation of the mercury exposure and lasted for more than nine months following cessation of exposure to mercury exposure (Wakita, 1987).

Hypertensive effects of mercury toxicity can be due to the action of mercury on the smooth muscle walls of the blood vessels. Vascular walls contain smooth muscle fibers that can constrict or relax, thereby influencing the flow of blood through the vessel. If large segments of the vascular tree constrict during the same time period, we find that blood pressure rises. The elevated blood pressure is due to the fact that there is a smaller intravascular volume available. As the space available to contain the blood supply (intravascular volume) diminishes, pressure increases.

Additional metal toxicities may occur concomitantly with mercury exposure. Cadmium, nickel, and lead are perhaps the most common co-existing heavy metal toxicity that can influence blood pressure and cardiovascular disease. A toxicology study published in 1990 revealed that mercury, cadmium, nickel and other heavy metal toxins raise blood pressure via vasoconstriction (Evans and

Weingarten). The study, done on dogfish sharks, showed that exposure to various levels of mercury and other heavy metal toxins caused varying degrees of constriction of blood vessels due to stimulation of the smooth muscle walls of the blood vessels. Lower doses of mercury and toxic metal exposure resulted in lesser constrictive effects whereas higher doses of mercury led to more significant vasoconstriction. It is significant that their research indicates that nickel should be investigated more thoroughly by scientists due to its potential role in causing hypertension.

A scientific study of the effects of organic methylmercury in rats showed that exposure to the potent toxicant increased blood pressure (Wakita). The study showed that when rats were exposed to organic mercury their systolic blood pressure increased significantly. Specifically, acute doses of methylmercury chloride at 5 mg/kg body weight per day and chronic doses of methylmercury chloride at 0.5 mg per kilogram of body weight per day led to increased systolic blood pressure in the animals when measured by the tail-cuff method. This systolic blood pressure increase was found to be long lasting in the chronically treated rats.

The trace mineral selenium has been shown in at least one study to counteract the toxicities of heavy metals including mercury (Whanger). The study showed that the element selenium, which is found in natural foods and is included within some multi-vitamin and mineral supplements, can counteract toxicities due to heavy metals. The metals tested and found to be beneficially affected by selenium included both inorganic and organic mercury, cadmium, thallium, and silver. The study also showed that vitamin E has similar effects to counteract heavy metal toxicities in the same list of metals. Specifically, vitamin E was found to significantly alter methylmercury toxicity. Vitamin E was more effective against lead than was selenium. Although this study was primarily done to elucidate the effects of selenium and vitamin E to counteract carcinogenic effects of heavy metals, it seems reasonable to extrapolate that cardiovascular toxicities may also be diminished by the use of selenium and vitamin E. Certainly, more studies are warranted to ascertain the validity of that idea.

Heavy metal toxicity, including mercury in both its organic and inorganic forms, have been found to cause carcinogenesis and mutagenesis. Carcinogenesis and mutagenesis are terms describing the transformation on normal genes and cells into cancers. In the selenium and vitamin E trial of Whanger, it was found that cadmium, inorganic mercury, methylmercury, thallium, and silver have less of an effect to produce tumors in animals when selenium treatment was given. Vitamin E was even more powerful to decrease the carcinogenic effects of methylmercury. Another study done long-term showed an increase of lung cancer rates among those exposed to significant mercury vapor inhalation (Ellingsen et. al.). This long-term study of workers exposed to industrial levels of mercury vapor from the years 1953 through 1989 showed that there was an increase in lung cancer among those workers exposed. The study failed to show an increase in cancer rates involving the kidneys, central nervous system, or brain.

Leukemia rates were found to be elevated among farmers who used mercury containing fungicides (Janicki, 1987). The study showed that the farmers who used mercury containing fungicides had elevated hair levels of mercury and had increased incidents of acute leukemia as compared to farmers not using fungicides. The same study also showed that cattle exposed to seeds coated with mercury-containing substances had an increased incidence of leukemia.

Other older studies, such as the 1950 study of Fitzhugh (Fitzhugh, 1950), failed to show increased incidences of cancers in animals experimentally exposed to mercury-containing substances. The Fitzhugh study exposed male and female rats to mercuric acetate as a dietary substance. Rats and others exposed to dietary phenylmercuric acetate did not exhibit cancerous or pre-cancerous lesions. However, noncancerous renal (kidney) lesions were seen in the animals. It should be noted that the study size was limited to small groups and that survival data was not reported. It is also interesting to note than an unspecified number of the animals died from pneumonia. These inconsistencies in the study may have reduced the sensitivity of the study to detect cancerous lesions or carcinogenic effects.

Another way to detect possible carcinogenic effects of mercury toxicity is to study the genotoxic effects of the metal. Genotoxicity refers to chromosomal or genetic abnormalities that occur upon exposure to certain toxic substances including mercury. Abnormalities in the structure of chromosomes have been reported in human white blood cells "lymphocytes." Chromosomal abnormalities in humans exposed to mercury vapors were found by Popescu (Popescu, 1989) and Verschaeve (Verschaeve, 1976). Aneuploidy gene formations among the workers exposed to mercury were found to be significantly increased in Verschaeve's study. Human lymphocyte analysis has shown that dietary mercury exposure, and increased blood mercury levels, are positively correlated with greater frequencies of chromosome abnormalities including aneuploidy and sister chromatid exchanges (Skerfving, 1974) (Wulf, 1986).

The two aforementioned dietary studies were done on subjects who consumed contaminated fish or seal meat and who exhibited high blood levels of mercury upon testing. These studies may not have sufficiently eliminated the possibility of concurrent and confounding genotoxic agent exposure. Mercuric acetate was found to produce DNA damage in cells that were grown "in vitro" (under laboratory conditions) (Williams et. al., 1987).

Another in vitro study showed genotoxic effects of mercuric chloride to induce mutations in mouse lymphoma cells (Oberly, 1982). Further in vitro studies showed chromosomal abnormalities in rat fibroblasts and Chinese hamster ovary cells (Rozalski and Wierzbicki, 1982) (Cantoni, et. al., 1984) (Christie, 1984, 1985). Studies showed that mercuric chloride can cause chromosomal damage by single strand DNA breakages (Cantoni and Costa, 1983). The same study found that the effects of mercuric chloride to damaged DNA is enhanced by the concurrent inhibitory effect of mercuric chloride on DNA repair mechanisms. Genotoxicity studies in live animals, also called "in vivo" studies, have also shown detrimental chromosomal mutations and abnormalities. Mercuric chloride was found to produce lethal mutations in rat cells (Zasukhina, 1983) and was found to bind chromatin in mouse liver cells (Bryan, et. al., 1974). Cats exposed to methylmercuric chloride were found to develop chromosomal aberrations (Miller et. al., 1979).

In order for mercury to cause toxic effects within the body it must be absorbed into the body and then be distributed into various tissues and organs where it may do damage. Responding to the toxicity, the body is continually working to eliminate the mercury. The absorption, distribution and elimination of a toxic metal such as mercury is described by a science called "toxicokinetics." The toxicokinetics of mercury has been studied in both humans and animals. However, due to the toxic nature of mercury, the majority of the work, and therefore knowledge about mercury toxicokinetics, comes from animal studies. We are able to extrapolate from animal studies the information lacking from human studies and thereby develop a fuller picture of the toxicokinetics of mercury within the body.

Let us first discuss the absorption of mercury. Metallic mercury fumes are easily diffused into the blood stream upon inhalation and are also lipophilic. The lipophilic nature of metallic mercury vapor allows it to diffuse through lipid or fatty areas of the body. Perhaps the most clinically important of the fatty tissues through which this type of mercury can diffuse are the nervous tissues of the body, such as the brain, spinal cord and peripheral nerves (Friberg and Vostal, 1972). A large percentage of inhaled mercury vapor is retained within the tissues of the human body. This has been measured at approximately 74 to 80 percent (Hursh et. al., 1976). The absorption of inorganic mercury through oral exposure indicates a limited absorption. An absorption of approximately 0.10 percent for metallic mercury taken by oral route has been estimated (Friberg and Norbery, 1973). However, long-term exposures may result in significant tissue burdens of the toxin.

Studies of an individual who was subjected to chronic ingestion of a mercury containing laxative showed high levels of mercury in various organ tissues, particularly the kidney (Weiss, et al., 1973). Organic compounds of mercury are much more easily absorbed by humans. Approximately 95 percent of methylmercuric nitrate is absorbed (Aberg et. al., 1969).

To summarize, the absorption rate of mercury depends upon the type of mercury involved. Metallic mercury vapors and mercury in other inorganic forms are much less readily absorbed into the human body. However, constant or chronic exposure to inorganic mercury can lead to elevated levels of the toxic metal within the tissues of the human body. Conversely, organic mercury compounds are readily absorbable and can easily lead to toxic levels in the human body.

How is the absorbed mercury distributed throughout the body? Mercury in its metallic form distributes readily to all tissues. Single-dose exposures of mercury have been found to cause peak levels to be reached within the tissues of the human body within twenty-four hours, with the exception of the central nervous system tissues, where the peak levels are not achieved until forty-eight or seventy-two hours after exposure. Studies have indicated that the kidney is the organ with the highest accumulation of inorganic mercury (Hursh et. al., 1976).

Organic compounds of mercury such as methylmercury also readily distribute throughout the tissues of the human body following absorption. Methylmercury is particularly mobile, easily penetrating into tissues which are predominately either watery (aqueous) or fatty (lipid). Levels of methylmercury are, therefore, a reliable indicator of the concentrations of methylmercury within other tissues (Nordberg, 1976) (Aberg, et. al., 1969). Once again, however, the highest levels, on the average, are still to be found within the kidneys.

In addition to blood being a good indicator of methylmercury levels in other tissues, hair can also be used. Organic mercury accumulates readily in hair and it is generally accepted that the relative mercury concentration in hair is directly proportional to mercury concentrations in the blood and therefore in other tissues (Phelps, et. al., 1980).

To summarize, the distribution of mercury within the tissues of the human body is widespread, with the kidneys becoming somewhat more affected than other types of tissues or organs. Organic compounds of mercury are even more readily distributed throughout the tissues of the body than are inorganic forms of mercury, although both are readily distributed throughout the human body.

As mercury is distributed throughout the tissues of the human body, it is metabolized in certain ways and then eliminated through various routes of excretion. Metallic mercury and other inorganic forms of mercury are rapidly oxidized by red blood cells, liver, and within the lung. This oxidation appears to primarily occur via the endogenous enzyme called "hydrogen peroxide catalase" (Halbach and Clarkson, 1978).

The oxidation of mercury is inhibited by ethanol. Ethanol alcohol is thought to inhibit the oxidation of mercury because it competes for oxidation with the same enzyme that oxidizes mercury. It is therefore a competitive substrate for hydrogen peroxide catalase. The presence of ethanol in humans at the time of exposure to metallic mercury thereby decreases the tissue concentrations of mercury (Hursh et. al., 1980). Organic mercury compounds, such as methylmercury, are also eventually oxidized. It is believed that methylmercury is first converted into inorganic mercury before the oxidation occurs.

Following metabolism, mercury in either its inorganic or organic form may be excreted through the body. This excretion occurs primarily through the urine or feces. Additionally, some mercury is eliminated through exhalation during breathing, or is excreted through the bile, sweat, or saliva (Lovejoy, et. al. 1974). As acute exposure to new doses of mercury metal occur, the body quickly eliminates the majority of the mercury. However, approximately fifteen percent of an acute dose will persist and remain for a protracted period, particularly in the brain and nervous system (Takhata, et. al. 1970).

Organic forms of mercury are particularly likely to be eliminated in feces. It appears that a significant percent of all organic mercury that initially enters the lower human digestive tract is there converted into inorganic forms by microorganisms that are found within the intestine (Yakanura, et. al., 1977).

Half-life of a substance is the time required for one half of a given dose to be eliminated or detoxified. Studies indicate that the half-life of methylmercury, the most common type of organic mercury to which humans are exposed, is approximately seventy to seventy-nine days within the tissues of the body (Aberg, et. al., 1969).

Physicians can assist the body in its attempts to detoxify and eliminate toxic mercury. Several drugs have been developed which can be used to entrap and remove mercury from the tissues of the body. Chelating drugs are those which entrap toxic metals and thereby allow them to be removed from the body via the urine or feces (Goodman & Gilman, fifth edition).

Commonly used chelating agents for toxic metals include DMPS, DMSA, and EDTA. DMPS (2, 3 dimercaptopropane-1-sulphonate) was developed in China and first used therapeutically in the Soviet Union as a treatment for workers exposed to toxic heavy metals. It works by chelating organic and inorganic mercury, lead, cadmium, arsenic, silver and tin.

DMPS has a very high affinity for mercury, but is unable to cross the blood brain barrier. It is therefore useful for non-CNS, systemic, mercury detoxification. DMSA (meso-2, 3-dimercaptosuccinic acid) also has a very high affinity for binding mercury. It can cross the blood brain barrier and therefore is more useful for central nervous system detoxification of mercury. Both DMSA and DMPS are eliminated primarily via the urine if given intravenously and also partially via the feces if given orally.

EDTA (ethylene diamine tetracetic acid) is best known as a chelator of lead. It has been used for decades in the treatment of "plumbism" in children. It has a lesser affinity for mercury than does DMPS or DMSA. EDTA is very poorly absorbed orally and is usually administered intravenously.

Certain vitamins, minerals and nutritional metals are antagonistic to the toxic effects of mercury and can be employed as medical treatment. Vitamins E and D, some of the most potent antioxidant vitamins, are among this group. Similarly, sulfur-containing compounds, also known as "thiols," which are known to be potent antioxidant organic materials, are also antagonistic to mercury toxicity. Copper, zinc, and selenium, all well-known metallic mineral nutritional supplements, have also been found to diminish the toxic effects of mercury (Ganther, 1980; Magos and Webb, 1976; Hansen and Kristensen, 1980; Welsh, 1979). Selenium, in the form known as sodium selenite, is sometimes used to counteract acute mercury poisoning in emergencies (Mengel and Karlog, 1980).

Case History 1:

Our patient, D.D., a gentleman in his late seventies, was brought to our clinic by his wife. Mr. D. had been recently diagnosed as suffering from Alzheimer's Disease by a neurologist. His symptoms of forgetfulness had begun several years earlier, and had progressed to the point that he was unable to drive his car, remember anything, or hold a sensible conversation. "Aricept" had not helped his condition.

At the time I initially interviewed him and his wife, he was not responsive to me in any meaningful way. Talking to him was essentially like talking to the wall. He gave no indication of realizing that he was being engaged in conversation, and there was no sign of any response from him, either verbal or non-verbal.

Testing revealed him to have several toxic metals with a prominence of aluminum. We began him on a treatment regimen consisting of intravenous therapy for removal of toxic metals, and herbs, which he took as oral supplements to stimulate the brain cells. Several weeks of treatment went by, and D.D. gradually came back into our world. One day, while talking to both he and his wife, I noticed he seemed to be making some mental calculations. His wife had just asked me how many more days until the treatments would be concluded, and I told her there were so many left, and that we were administering them at a certain rate per week. Before I could make the calculation of how many more weeks of treatment would therefore remain, Mr. D. interjected with the answer. And he was correct. An unbelievable improvement in his condition had occurred!

Case History 2:

One of our patients, treated several years ago, demonstrates the lasting effects that can potentially be obtained by the elimination from the body of mercury and related toxic metals. P.M. was a 38-year old man four years ago when first tested for metal toxins. He originally complained of a life-long history of severe seasonal allergies. Hair analysis testing revealed a very high level of systemic mercury, with lesser but substantial levels of cadmium and lead.

Intravenous treatments were administered to remove these toxic elements. He underwent a total of twelve such treatments, and after the eighth, we first noticed a dramatic reduction in symptoms.

Four years have now passed and he continues to report a lasting dramatic reduction in allergy symptoms. P.M. reports approximately an eighty-percent reduction in seasonal allergy symptoms. Whereas he initially described himself as highly allergic with a severe problem, he now considers himself to have reverted to a condition of mild seasonal allergies. Happily, he states that he has not needed to use any allergy medication at all since his treatments four years ago.

How to Test and Remove Mercury

If you are concerned about the possibility of mercury toxicity in your system, there are two common ways to test for it. You can request that your family doctor orders a blood test for the metal, or that he orders a hair analysis for toxic elements. Certain labs are more proficient at performing a quality test than are others, so your doctor should do some research to find a high quality testing source for the hair analysis.

The most powerful treatments for mercury are intravenous therapies administered by a medical doctor. However, on your own, you may possibly reduce your systemic burden of the toxic metal. Garlic, a common herb, is the easiest and most readily available detoxification food substance. Parsley and alfalfa also seem to be very useful in such cleansing. Supplements that can be found in most health food stores may contain these and other substances to cleanse and detoxify the body of mercury and other harmful metals. Some of these supplements are called "oral chelators," even though they may not actually "chelate" anything, but in any event, many of them do seem to help to cleanse the body of toxins.

References

EPA 1971a. "National Emissions Standards For Hazardous Air Pollutants, List Of Pollutants And Applicability Of Part 61," *Federal Register* 36:5931.

EPA 1971b. *Water Quality Criteria Data Book.* United States Environmental Protection Agency, Washington, D.C.

EPA 1975. "National Emissions Standards For Hazardous Air Pollutants, Emissions Standard," *Federal Register* 40:59570.

EPA 1984. "Mercury Health Effects Update; Health Tissue Assessment; Final Report," United States Environmental Protection Agency, Washington, DC; EPA 600 8-84-019F.

EPA 1987. "Health Advisory For Mercury," United States Environmental Protection Agency, Washington, DC.

Enwonwu CO, 1987. "Potential Health Hazards Of Uses Of Mercury In Dentistry; Critical Review Of The Literature," Environmental Research 42:257-274.

Weiss, G., 1986. *Hazardous Chemicals Data Book.* Second edition. Noyes Data Corp.

PHS, 1989. "Toxicological Profile For Mercury," Agency for Toxic Substances and Disease Registry, United States Public Health Service.

Campbell, J., 1948. "Acute Mercurial Poisoning By Inhalation Of Metallic Vapor In An Infant." Canadian Medical Association J 58:72-75.

Matthes, F., Kirschner, R., Yow, M., et. al. 1958. "Acute Poisoning Associated With Inhalation Of Mercury Vapor. Report of four cases," *Pediatrics* 22:675-688.

Teng, C., Brennan, J., 1959. "Acute Mercury Vapor Poisoning. A Report Of Four Cases With Radiographic And Pathologic Correlation," *Radiologon* 73:354-361.

Tennant, R., Johnston, H., Wells, J., 1961. "Acute Bilateral Pneumonitis Associated With The Inhalation Of Mercury Vapor. A report of five cases," *Conn Med* 25:106-109.

Hill, W., 1943. "A Report On Two Deaths From Exposure To The Fumes Of A Di-ethyl Mercury," *Canadian Journal of Public Health* 34:158-160.

Campbell, J., 1948. "Acute Mercurial Poisoning By Inhalation Of Metallic Vapor In An Infant," *Canadian Medical Association* 58:72-75.

Hallee, T.J., 1969. "Diffuse Lung Disease Caused By Inhalation Of Mercury Vapor," *Am Rev Resp* Dis 99:430-436.

Wakita, Y. "Hypertension Induced By Methylmercury In Rats," *Toxicology of Applied Pharmacology*, 1987 Jun 15th; 89-1: 144-147.

Evans, D. H., Weingarten, K. "The Effect Of Cadmium And Other Metals On Vascular Smooth Muscle Of The Dogfish Shark, Squalus Acanthias," *Toxicology*, 1990 Apr 30th; 61-3: 275-281.

Wakita, Y., "Hypertension Induced By Methylmercury In Rats," *Toxicology of Applied Pharmacology*, 1987 Jun 15th; 89-1:144-147.

Whanger, P.D., "Selenium In The Treatment Of Heavy Metal Poisoning And Chemical Carcinolenesis," *Journal of Trace Element Electrolytes in Health and Disease*; 1992 Dec; 6-4: 209-221.

Ellingsen, D. G., Andersen, A., Nordhagen, H. P., Efskind, J., Kjuus, H., "Incidents Of Cancer And Mortality Among Workers Exposed To Mercury Vapor In The Norwegian Chloralkali Industry," *British Journal of Industrial Medicine* 1993 Oct; 50-10: 875-880.

Janicki, K., Dobrowolski, J., Drasnicki, K., 1987. "Correlation Between Contamination Of The Rural Environment With Mercury And Occurrence Of Leukemia In Men And Cattle," *Chemosphere* 16:253-257.

Fitzhugh, O.G., Nelson, A.A., Laug, E.P., et. al. 1950. "Chronic Oral Toxicities Of Mercuri-phenyl and Mercuric salts," *Arch Ind Hyg Occup Med* 2:433-442.

Popescu, H. I., Negru, L., Lancranjan, I., 1979. "Chromosome Aberrations Induced By Occupational Exposure To Mercury," *Arch Environmental Health* 34:461-463.

Verschaeve, L., Kirsch-Volders, M., Hens, L., et. al. 1978. "Chromosome Distribution Induced By Occupationally Low Mercury Exposure," Environ Res 12:306-316.

Skerfving, S., 1974. "Methylmercury exposure, Mercury Levels In Blood And Hair, And Health Status In Swedes Consuming Contaminated Fish," *Toxicology* 2:3-23.

Wulf, H. C., Kromann, N., Kousgaard, N., et. al. "1986 Sister Chromatid Exchange (SCE) In Greenlandic Eskimos. Dose-response Relationship Between SCE And Seal Diet, Smoking, nd Blood Cadmium And Mercury Concentrations," *Sci-Total Environ* 48:81-94.

Williams, M.V., Winters, T., Waddel, K.S., 1987. "In Vivo Effects Of Mercury (II) On Deoxyuridine Triphosphate Nucleotidohydrolase, DNA Plymerase (Alpha, Beta), And Uracil-DNA Glycosylase Activities In Cultured Human Cells: Relationship To DNA Damage, DNA Repair, And Cytotoxicity," *Mol. Pharmacol* 31:200-207.

Oberly, T.J., Piper, C.E., McDonald, D.S., 1982. "Mutagenicity Of Metal Salts In The L5178Y Mouse Lymphoma Assay," *Journal of Toxicology and Environmental Health* 9:367-376.

Rozalski, M., Wierzbicki, R., 1983. "Effect Of Mercuric Chloride On Cultured Rat Fibroblasts: Survival, Protein Biosynthesis And Binding Of Mercury To Chromatin," *Biochemical Pharmacology* 32:2124-2126.

Cantoni, O, Christie, N.T., Swann, A., et. al. 1984. "Mechanism Of Hgcl2 Cytotoxicity In Cultured Mammalian Cells," *Mol Pharmacology* 26:360-368.

Christie, N. T., Cantoni, O., Sugiyama, M., et. al. 1985. "Differences In The Effects Of Hg(II) On DNA Repair Induced In Chinese Hamster Ovary Cells By Ultraviolet Or X-Rays," *Mol Pharmacol* 29:173-178.

Cantoni, O., Costa, M. 1983. "Correlation Of DNA Strand Breaks And Their Repair With Cell Survival Following Acute Exposure To Mercury (II) And X-Rays," *Mol Pharmacology* 24:84-89.

Zasukhina, G. D., Vasilyeva, I. M., Sdirkova, N. I., et. al. 1983. "Mutagenic Effect Of Thallium And Mercury Salts On Rodent Cells With Different Repair Activities," *Mutat Res* 124:163-173.

Bryan, S. E., Guy A. L., Hardy, K. J. 1974. "Metal Constituents Of Chromatin Interaction Of Mercury In Vivo," *Biochem* 13:313-319.

Miller, C. T., Zawidska, Z., Nagy, E., et. al. 1979. "Indicators Of Genetic Toxicity In Leukocytes And Granulocytic Precursors After Chronic Methylmercury Ingestion By Cats," *Bulletin of Environmental Contamination and Toxicology* 21:296-303.

Friberg, L., Vostal, J., ed. 1972. "Mercury In The Environment: A Toxicological And Epidemiological Appraisal," Cleveland, OH: CRC Press.

Hursh, J. B., Clarkson, T. W., Cherian, M. G., et. al. 1976. "Clearance Of Mercury (Hg-197, Hg-203) Vapor Inhaled By Human Subjects," *Arch Environ Health* 31:302-309.

Fribery, L., Nordberg, F., 1973. "Inorganic Mercury - A Toxicological And Epidemiological Appraisal," In Miller, M. W., Clarkson, T. W., ed. *Mercury, Mercurials And Mercaptans,* Springfield, IL; Charles C. Thinasm 5-22.

Weiss, S. H., Wands, J. R., Yardley, J. H. 1973. "Demonstration By Electron Defraction Of Black Mercuric Sulfide (B-Hgs) In A Case Of Melanosis Coli And Black Kidneys Caused By Chronic Inorganic Mercury Poisoning," (Abstract). *Lab Invest*:401-402.

Aberg, B., Ekman, R., Falk, U. et. al. 1969. "Metabolism Of Methymercury (203Hg) Compounds In Man: Excretion And Distribution," *Arch Environ Health* 19:478-484.

Hursh, J. B., Clarkson, T. W., Cherian, M. G., et. al. 1976. "Clearance Of Mercury (Hg-197, Hg-203) Vapor Inhaled By Human Subjects," *Arch Environmental Health* 31:302-309.

Nordberg, G., ed. 1976. "Effects And Dose-Response Of Toxic Metals," *New York: Elsevier/North Holland Biomedical Press.*

Aberg, B., Ekman, R., Falk, U. et. al. 1969. "Metabolism Of Methylmercury (203Hg) Compounds In Man: Excretion And Distribution," *Arch Environmental Health* 19:478-484.

Phelps, R. W., Clarkson, T. W., Kershaw, T. G., et. al. 1980. "Interrelationships Of Blood And Hair Mercury Concentrations In A North American Population Exposed To Methylmercury," *Arch Environ Health* 35:161-168.

Halbach, S., Clarkson, T. W., 1978. "Enzymatic Oxidation Of Mercury Vapor By Erythrocytes," *Biochem Biophys Acta* 523:522-531.

Hursh, J. D., Greenwood, M. R., Clarkson, T. W., et. al. 1980. "The Effect Of Ethanol On The Fate Of Mercury Vapor Inhaled By Man," *Journal of Pharmacology Exp Ther* 214:520-527.

Lovejoy, H. B., Bell, Z. G., Vizena, T. R., 1974. "Mercury Exposure Evaluations And Their Correlation With Urine Mercury Excretion," *Journal of Occupational Medicine* 15:590.

Takahata, N., Hayashi, H., Watanabe, B., et al. 1970. "Accumulation Of Mercury In The Brains Of Two Autopsy Cases With Chronic Inorganic Mercury Poisoning," *Folia Psychiatric Neurology Japan* 24:59-69. Yakanura, et. al., 1977.

Aberg, B., Ekman, R., Falk, U., et. al. 1969. "Metabolism Of Methylmercury (203Hg) Compounds In Man: Excretion And Distribution," *Arch Environ Health* 19:478-484.

Goodman, L. S., Gilman, A., 1975. *The Pharmacologic Basis of Therapeutics* Fifth Edition, 912-921. MacMillan Publishing Co.

Ganther, H. E., 1980. "Interactions Of Vitamin E And Selenium With Mercury And Silver," *Academy of Science* 355:212-216.

Magos, L., Webb, M., 1976. "The Interaction Between Cadmium, Mercury, And Zinc Administered Subcultaneously In A Single Injection," *Arch Toxicology* 36:53-61.

Hansen, J. C., Kristensen, P., 1980. "On The Influence Of Zn On Hg/Se Interaction," *Arch Toxicology* 46:273-6.

Welsh, S. O., 1979. "The Protective Effect Of Vitamin E And N, N-Diphyenyl-P-Phenylenediamine Against Methylmercury Toxicity In The Rat," *Journal Nutr*:1973-81.

Mengel, H., Karlog, O., 1980. "Studies On The Interaction And Distribution Of Selenite, Mercuric, Methocyethyl Mercuric And Methylmercuric Chloride In Rats," *Acta Pharmacol* Toxicol II:25-31.

SECTION THREE
HORMONE REPLACEMENT THERAPY

Chapter Five
DHEA & HUMAN GROWTH HORMONE REPLACEMENT THERAPY

The Concept of Hormone Replacement Therapy

Our body secretes powerful chemicals to control the most vitally important physiologic processes. These chemicals control our physiology by acting as messengers and signalers. They travel via the bloodstream to cells, tissues, and organs, causing them to change their activity levels. Such messengers are known as "hormones."

Hormones such as human growth hormone (hGH), thyroid, and the sex hormones are the most powerful components of the "commands" that cause bodily development to progress from childhood to the completion of puberty. Once adulthood is reached, these hormones tend to decline, some precipitously. Declining hormonal command and control mechanisms are conjectured by many anti-aging experts to be a fundamental cause of the aging process.

The goal of hormone replacement therapy is to restore youthful or healthful levels of these critical hormones. Restoring the "hormones of youth" promotes within an adult body a more youthful physique and appearance by creating a more longevous and vigorous form and function of each cell, tissue, and organ. Rejuvenating the aged individual and repressing the aging process in younger people is today primarily achieved by the medical use of hormone replacement therapy.

Hormone replacement therapy is the single most powerful anti-aging therapy currently utilized by most physicians. However, hormone replacement therapy does not get to the root of the problem. Hormones decline with age primarily because one or more genetic, inherited, factors direct the body to make less of them. The ultimate hormonal deficiency treatment will be one that reprograms this genetic message that instructs the body to lessen the production of the hormone of youth. Telomerase therapy may soon serve to allow us to change the genetically inherited command to curtail production of the vital hormones. It is likely that the addition of telomerase treatment to the currently used regimen of hormone replacement therapy will yield dramatic anti-aging results.

Human Growth Hormone Replacement Therapy

Human Growth Hormone, also known as hGH, is the single most powerful hormone available to modern physicians as they seek to rejuvenate the aging human body. The restorative and rejuvenative effects of hGH replacement therapy are dramatic. They include, 1) loss of body fat and increases in lean muscle, 2) generalized increased strength, vigor, and energy, 3) more youthful appearance of the skin from restored structural integrity, elasticity, and lessened wrinkling, 4) improved sexual functioning, and 5) normalization of blood fats including cholesterol, HDL and triglycerides. The effects in some patients are so dramatic that they have been described as "aging reversal."

What is this remarkable hormone of youth and why do we eventually stop producing it after puberty? Human Growth Hormone is composed of poplypeptides (protein-like material) and is produced in the pituitary gland. The pituitary is an area of the brain known for its production of essential regulatory hormones. The pituitary releases growth hormone in various amounts on a periodic basis. This periodic release of hormone mostly occurs in a twice-daily pattern.

The quantity of growth hormone released is in relationship to the age of the individual, the exercise levels of the individual, whether the individual has low blood sugar or is in a fasting condition, whether trauma has occurred to the body, and other factors. It has also been noted that several types of dietary supplements and medications can increase growth hormone production and release by the pituitary gland.

Laboratory Testing for hGH Levels

How can we scientifically determine our hGH blood levels? Measurements of growth hormone can be taken through blood tests, which are commonly available from clinical testing laboratories. Testing for human growth hormone generally costs less than one hundred dollars ($100) and can give us a good indication of the amount of human growth hormone an individual's body is producing at the time.

Another way to measure levels is to test the blood concentration of IGF1, the immediate successor to growth hormone. When growth hormone is released into the body stream, it passes into the liver, which is thereby stimulated to produce IGF1. IGF1 is then released into the bloodstream and produces most of the physiological effects attributed to growth hormone. Measurement of IGF1 can, therefore, give us a good indication of growth hormone functional levels, and may be less expensive to test than growth hormone.

After testing the growth hormone level, you must interpret the significance of the reading. What level should it be? This varies for the individual. However, it is usually thought that a growth hormone level equal to that found in an individual in their late twenties may be optimal as a life extension therapy.

Due to potential side effects of high dose human growth hormone, it seems unreasonable to attempt to obtain an amount of human growth hormone in the body that is equal to that of an adolescent. Adolescents require very large amounts of human growth hormone for their bodily development. This very high level of human growth hormone may actually have more damaging effects than beneficial effects for older adults who are interested in life extension.

hGH Deficiency

Average blood levels of hGH change as we pass through the various stages of life. Changes in growth hormone levels likely have more profound effects upon the condition of our body than does any other single hormone. Large releases of hGH during childhood causes normal physical growth and development. Severe deficiencies in growth hormone during childhood may actually cause "dwarfism."

Children suffering from dwarfism have an inborn diminished ability of their body to produce growth hormone. After full maturity has been reached, growth hormone production generally declines. As a matter of fact, it declines steadily with age with the greatest decline occurring between puberty and early middle age. By the sixth through eighth decades of life its production is virtually nil.

Low levels of this hormone in adults produce what is known as "adult growth hormone deficiency syndrome." Adult human growth deficiency syndrome occurs when levels of the hormone drop low enough so that the body deteriorates as a result of the lack of tissue stimulation by the hormone. Although we all experience declining hGH levels as we age, different people in early, mid, and late adulthood show varying levels of growth hormone production by their bodies. Those with low levels will notice increased fat, lessened muscle mass, thin and sagging skin, lost energy, and diminished vitality.

Perhaps one of the most significant differentiating factors separating those adults who produce more growth hormone as opposed to those who produce less growth hormone, is how much and what kind of exercise they engage in. People who use their muscles to lift heavy weights or in strenuous exertion produce more growth hormone. Heavy physical exercise, in which there is at least a brief period of sustained significant muscular effort and work, causes the body to release human growth hormone as a response to that exercise.

Researchers are presently not sure whether the ability of the body to produce this exercise-induced human growth hormone release is maintained throughout life. Some researchers believe that in many individuals the ability to produce human growth hormone in response to exercise diminishes greatly or is completely lost after the third decade of life.

Heavy physical exercises that do not involve the muscles being put to great stress (as far as requiring the muscles to lift a heavy load or produce great force) usually do not result in a significant growth hormone release. In particular, jogging and other light aerobic exercises usually do not cause any of the muscles of the body to be exercised to such a peak output that the pituitary responds by producing significantly more growth hormone.

The exercise-induced production and release of growth hormone primarily occurs during the sleeping periods following exercise. The night's sleep after heavy exercise or even short daytime naps have been shown by scientists to be a period of time in which the pituitary gland of the brain releases its growth hormone to the body via the blood stream. Good sleep habits are thus important to human growth hormone release and production.

The exercise-induced growth hormone helps to maintain or, in some instances increase, the muscle mass and its accompanying bone mass. In adults, hGH functions not only to maintain muscle, but the strength and youthfulness of the skin as well. It is important in maintaining the health of the immune system and has been found to burn fat.

Those who have higher levels of human growth hormone in their systems are usually much leaner due to a lesser percentage of fat and a greater percentage of muscle, bone and lean body tissues than are those individuals who have low growth hormone levels. Part of this occurs from the metabolism or "burning" of fat. Metabolism of fat uses the fat as an energy source or as a building block to build the leaner tissues of the body. One of the hallmarks of the people who have adequate growth hormone levels is that they are leaner, more muscular and have a much smaller percentage of total body fat. Studies also suggest that they live longer, healthier, and more vigorous lives.

Natural hGH Inducers

In addition to exercise, several other natural factors can increase our production of hGH. Many vitamins and foods have been shown to cause the production and release of growth hormone in those individuals who still possess the capability to make the hormone.

Niacin is a commonly used vitamin supplement and component of food. Taken in doses much higher than that encountered in foods, it can have many beneficial health effects such as a marked natural growth hormone stimulator. Dosages from 200 to 1000 mg have been reported to raise growth hormone levels. Xanthinol nicotinate is perhaps the most powerful form of niacin to be used as a growth hormone releaser. Niacin is currently used by both conventional and alternative medical doctors to lower cholesterol and improve the blood fat profile. It can lower total cholesterol, raise HDL, and lower triglycerides. The most common doses used for modification of the blood lipid profile are 1000-3000 mg per day.

When using niacin, we must be aware of an interesting, usually harmless side effect that can occur. That side effect is "flushing," a tingling, warm sensation that occurs simultaneously with a redness of the skin. This reaction lasts for several minutes and is caused by a niacin-induced release of histamine by the body. Gradually increasing the dose of niacin over a period of time minimizes this effect. Sudden exposure of high doses of niacin to the body virtually assures that this side effect will occur.

Another group of natural food substances useful in stimulating production of adequate hGH are the amino acids. Amino acids are found in foods as components of dietary proteins. Proteins are abundant in such foods as fish, meat and beans. The amino acids yield superior results when taken on an empty stomach at bedtime, on an empty stomach prior to heavy exercise, or both. Amino acids, along with other natural hGH releasing substances, seem to work best in those individuals under fifty years of age.

OKG is a man-made nutrient consisting of the amino acid ornithine bound to alpha keto glutarate. Its complete name is "L-ornithine alpha-ketoglutarate." OKG is reportedly well tolerated in the required large doses. These dosages are in the range of ten or more grams and it is recommended that the product be taken with ample fluids.

Arginine, ornithine, glutamine, glycine, tryptophane and lysine are some of the best known amino acid hGH stimulators. They are found in foods, but can be presented to the body in greater concentration as a supplement. Of these, arginine and ornithine have been the most widely used of the amino acids for hGH stimulation if ingested as pure amino acids in doses of a few grams per day. This supplementation is most effective when taken at bedtime, particularly on an empty stomach. Arginine and ornithine can be obtained through health food stores and many drug stores.

Specifically, scientists have found that L-arginine in doses of 2 to 5 grams with L-ornithine in doses of 2 to 5 grams taken on an empty stomach at bedtime is appropriate for causing growth hormone release. Glutamine, taken in doses of 2 to 2.5 grams at bedtime can produce a significant growth hormone release. Similarly, the amino acid trytophan may also be beneficial for growth hormone production when taken in a dose of 1 to 2 grams at bedtime on an empty stomach (Tryptophan is available by prescription only).

Stimulation of hGH by the amino acid "ornithine" appears to be significantly more powerful than arginine if the same amounts of each are taken. It should also be noted that it is not necessary to take both of these amino acids together. Ornithine, arginine, glutamine, lysine or tryptophan, whether alone or in combination, usually increase human growth hormone production in individuals who have the capability to have that production stimulated.

An additional benefit of arginine and ornithine supplementation is a stronger immune system with greater response and greater effectiveness against viruses, tumor cells and bacteria. Both amino acids, arginine and ornithine, have also been found to promote healing of wounds, and in some animals, to stimulate the regeneration of certain organs. These amino acids work best when taken at bedtime, just prior to exercise, or both, and on an empty stomach. When using these amino acids, it is wise to begin with lower dosages and slowly increase. Many people will develop gas or diarrhea if they take doses that are too large on an empty stomach.

A combination of two or more of the aforementioned amino acids may yield even more dramatic effects than using any one alone. Perhaps the most famous combination is that of arginine and ornithine. In addition to the dosages previously described, some advocate (see Pearson & Shaw in their book entitled *Life Extension, a Practical Scientific Approach*) arginine in doses of 5 to 10 grams and ornithine in doses of 2.5 to 5 grams (care should be taken to minimize bowel side effects by raising the dosage levels slowly over several weeks). Additionally, combinations of 2 grams of arginine, 2 grams of ornithine, 1 to 2 grams of lysine, and 1 to 2 grams of glutamine have been used.

Another strategy is to initially take 1 gram of arginine, 1 gram of glutamine, and 1 gram of lysine. The dosages of each of these three amino acids can then be raised slowly by 1 gram a week or as tolerated, until a dosage of 5 grams for each amino acid is reached. It should be noted that some combinations of amino acids might raise insulin levels as well as growth hormone levels. This effect is most commonly noted in the initial few months of treatment and may subside as the body becomes leaner under the influence of growth hormone. Individuals known to have blood sugar problems such as hypoglycemia (low blood sugar), hyperglycemia (elevated blood sugar), or diabetes mellitus (persistent substantial elevations of blood glucose) are best advised to use the amino acids only under the care of a physician, and to monitor their blood sugar levels closely.

Drugs That Improve hGH Levels

Several medications, most of them well known anti-aging drugs known for benefits other than growth hormone stimulation, have also been found useful to induce our production of the hGH. If any of these drugs are being used for other anti-aging purposes, we should check growth hormone levels before adding additional growth hormone stimulators such as amino acids. Most of these drugs have an additive or "synergistic" effect when used together with amino acids. The synergistic effect yields a greater growth hormone release than either drug or amino acid used singly.

One such natural growth hormone releaser is thyroid hormone. It has dramatic effects on the tissues and the functioning of the body in general. Thyroid hormone has been found to produce some of these effects by inducing an enhanced production of growth hormone by the body. Therefore, any use of thyroid hormone will likely have a beneficial effect on the production of human growth hormone.

Hydergine is a medical drug used as a growth hormone stimulant. It is generally well tolerated and has a significant history of clinical use. Additionally, it can have significant anti-aging actions apart from inducing growth hormone production. Clinical studies have shown that this broadly beneficial drug releases significant amounts of growth hormone in elderly test subjects. Suggested dosage for growth hormone release is in the six milligram per day range. This drug is generally well tolerated, but may cause bradycardia (slowed heart rate), rashes, nausea, drowsiness, and, when used with caffeine, can cause headaches or insomnia.

Bromocriptine, also known as the prescription drug Parlodel, is a medication that has been found to enhance growth hormone production by the body. Bromocriptine has many anti-aging effects including normalizing growth hormone output. Normalizing means that Bromocriptine can increase human growth hormone production when it is too low and diminish it if it is too high. Growth hormone may be overproduced in certain disease states exemplified by the disease "acromegaly."

This drug also appears to be able to reset some of the aging clocks found within the cells of the brain. Scientists have shown that it can reinstate menstrual cycles in some women who have gone through menopause. It can do the same thing in experimental animals, returning them to a youthful reproductive condition. This may be an important drawback to its use in human females, as older women who become pregnant have a much higher rate of genetic and birth defects among their offspring. Bromocriptine is a prescription medicine and should only be taken when prescribed and monitored by a licensed medical doctor.

Bromocriptine decreases the production of prolactin hormone in the body. Prolactin can act as an aging facilitator. As with L-Dopa, individuals who are schizophrenic should be given bromocriptine with great caution since it can exacerbate some of their symptomatology. It should also be used with caution if the patient has poorly controlled high blood pressure. It is interesting that hypotension, which is a word describing abnormally low blood pressure, can also occur and cause adverse symptoms in users of bromocriptine. Problematic side effects of bromocriptine, particularly among those who are new users of the drug, can include dizziness in eight to sixteen percent of people, drowsiness in eight percent of people, and fainting or feeling faint in approximately eight percent of people. Therefore, bromocriptine should be used with caution in people with hypertension and they should be monitored for periods of hypotension or low blood pressure.

In addition, users of bromocriptine should be aware of the significant percentages of individuals who may suffer from dizziness, drowsiness, faintness or fainting. Bromocriptine therapy is to gradually increase the daily dosage until the therapeutic response desired is achieved. Initial dosages of a half or whole 2.5 mg Parlodil snap tab is given daily. Therapeutic ranges are commonly found in the 5 to 7.5 mg per day area. Lower doses have also proven effective.

L-Dopa is a prescription drug made of an amino acid that has many effects on the brain. One effect is to help stimulate growth hormone production by the pituitary gland. It should be noted that its most famous use is in the treatment of Parkinson's Disease. Scientific experiments with L-Dopa show that elderly rats are able to regain much of their physical ability and after their treatment with L-Dopa, can swim as well as young adult rats. L-Dopa is also a potent antioxidant and can have beneficial effects on neurotransmitters in the brain that help control fine motor coordination and muscular movement.

Caution should be used when taking L-Dopa since it can cause harmful side effects. L-Dopa should only be taken under a doctor's prescription and under careful supervision. L-Dopa dose levels that have been found beneficial for increasing growth hormone production by the body are in the ¼ to ½ gram per day range. Once again, this prescription medicine is most beneficial when it is taken at bedtime on an empty stomach.

It should be noted that in some individuals, especially those who have Parkinson's disease and are taking L-Dopa for treatment of that disease, high doses of supplemental vitamin B-6 can lead to significant adverse interactions with the L-Dopa. It is thought that very high levels of vitamin B-6 in combination with L-Dopa can convert the L-Dopa to Dopamine outside the brain and at a level where it interferes with the body's functioning and worsens Parkinson's symptoms.

Some scientists have come to the conclusion that a combination of all four of these growth hormone releasing amino acids has an even greater effect than the use of any one alone. So it may be that by using a combination of tryptophan, L-Ornithine, L-Arginine, and L-Dopa at bedtime on an empty stomach could have a most dramatic effect on improving growth hormone production by the body.

Some scientists believe that this growth hormone production can be so great, even in people over age 60, that it can return their growth hormone level to that of a teenager. Since causing this much growth hormone production may not be most beneficial in anti-aging therapy, measuring the levels of human growth hormone after treatment and during treatment are an important part of any sensible anti-aging program.

Precautions for the use of L-Dopa include using it with high doses of supplemental vitamin B-6, in those people suffering with Parkinson's Disease, people with schizophrenia, and people with malignant melanoma cancers. Schizophrenics may be worsened by the use of growth hormone releaser including L-Dopa, arginine and ornithine and the growth hormone releasing drug bromocriptine. People with the cancer known as "malignant melanoma" should also be cautious. They should avoid the use of L-Dopa supplementation.

A new and potentially powerful growth hormone releaser is GHB or Gamma Hydroxy Butyrate. Scientists at the University of Chicago have discovered that GHB induced deep sleep acts to release human growth hormone from the brain in older test subjects. Although GHB is not presently available in the United States, it promises to one day become a potent addition to the medical treatment of aging.

Vasopressin or "ADH", which has the full scientific name "anti-diuretic hormone," is a hormonal substance produced by the body to control many physiologic responses. Vasopressin hormone is secreted by the posterior portion of the pituitary gland, which is found deep within the brain. Some of its physiologic functions include helping the body retain fluids, helping to increase the blood pressure when necessary, and causing muscles of the intestine, uterus and blood vessels to contract appropriately. As an anti-aging medication, Vasopressin has been found to have marked benefits on memory and learning functions in human beings. Most particularly, it has been found to be a growth hormone medication.

In addition to helping the posterior pituitary produce and release growth hormone, it also causes the production of another hormone called "beta lipotropin," which is a hormone causing fat to be metabolized and burned. Vasopressin is available as a prescription drug known as Diapid, a nasal spray medication. This allows the body to absorb the vasopressin through the mucus membrane of the nose.

Medical Hormones for Injection

Biologically active human growth hormone is now available for treatment of those diagnosed with "adult human growth hormone deficiency syndrome," which is caused by the normal, gradual reduction in our body's production of this "hormone of youth." This human growth hormone is manufactured via "biotechnology" in which human genes are transplanted (recombinant DNA) into microorganisms, which then make large amounts of the hormone.

Scientists believe lack of growth hormone to be a major causative factor in the initiation and progression of aging. Replacement of human growth hormone with mass produced replacement hormone can have dramatic rejuvenative effects on the body. These "aging reversal effects" can include, 1) loss of body fat and increases in lean muscle, 2) generalized increased strength, vigor, and energy, 3) improved strength and appearance of the skin with lessened wrinkling, 4) improved sexual functioning, and 5) improvements in blood fats. Companies currently manufacturing recombinant DNA produced hGH are Serrano Corp., Eli Lilly Corp., and Genentec Corporation. Currently, the hormone is available as long and short-acting injections.

Thyroid and Adrenal Hormone Replacement

The thyroid and adrenal are glands in the body which produce essential regulatory hormone substances. Hormones are substances produced by the body which are released from the site of production and are transferred, commonly through the blood, to various organs or tissues. They effect functions and activities of the target tissues and organs. Hormones commonly act to increase an organ or tissue's level of function. The "target" organ may then produce more of the products it manufactures, or increase the other activities that it performs for the body as a whole. In contrast, some hormones act in an opposite way so that they decrease the production from the target organ.

Thyroid hormone is perhaps the body's single most important regulatory hormone. It affects virtually every tissue and cell of the body and helps to set its level of metabolism functioning and reproduction. Thyroid hormone is the primary hormone that sets our body's basal metabolic rate. The basal metabolic rate "BMR" is the basic rate of cellular metabolism of the body at rest. Our cells metabolize energy by an oxidation-reduction reaction of nutrients and oxygen. Thyroid hormone helps to set this rate of metabolism. Therefore, its effects are profound, for both health and longevity.

Why is it that a hypothyroid "low functioning" condition would act as a factor which worsens and speeds the aging process? It may well be that thyroid hormone acts to fine-tune the body's functions and processes to operate at the most efficient level. This efficiency may result in less wear and tear on the cells of the body. Less abnormal metabolic byproducts may build up, and all of the tissues (including the circulatory tissues of the body) may function in ways that allow the body's natural anti-aging control mechanisms to function at their maximum rate. On the contrary, abnormally elevated functions, or hyperthyroidism, may cause rapid aging just as the hypothyroid state does. Hyperthyroidism may do this by once again throwing the body's metabolic functioning out of the efficient range by turning it up too high. Therefore, it is prudent to achieve a proper thyroid functioning as a crucial life extension therapy.

Adrenal hormones are produced by the adrenal gland and also act as very important, widespread, regulatory substances for the body. Adrenal hormones also help to set the basic energy levels of the body and its metabolic rate. However, thyroid seems to be by far the most potent hormone for setting the body's basic rate for metabolism and energy.

How can we tell if the adrenal or thyroid glands and their hormones are functioning properly? One way is to recognize the symptoms of poor thyroid and adrenal functioning. It is very common for both adrenal and thyroid function to be lacking as we age. This can result in a feeling of sluggishness, tiredness, and generalized lack of energy. As the problem becomes more severe, more profound symptoms of low thyroid or adrenal function can occur. Thickening of the skin, puffiness of the tissues, fluid retention, poor digestion with constipation, poor tolerance to emotional stress, decreased memory, diminished mental functioning and hair loss can ensue.

Aside from recognizing some of the signs and symptoms of low thyroid function, how can we determine more specifically whether there is a need for thyroid hormone replacement? One of the most common means used in conventional medicine is a blood test measuring the hormonal production of the thyroid gland. Measurements of total thyroid, T-3, T-4 and thyroid stimulating hormone (TSH) are gathered to determine whether a state of normal thyroid, hypothyroid, or hyperthyroid exists. Hypothyroid means low thyroid production or function. Hyperthyroid is just the opposite, and connotes excessive thyroid hormone release.

The blood tests for hypo or hyperthyroidism are sometimes not directly correlated with a person's symptomotology. Therefore, many alternative healthcare physicians believe that normal, and especially low normal thyroid blood test readings, do not necessarily mean that the person has normal thyroid hormone functioning.

Perhaps a more beneficial test is the early morning temperature test. To perform this test, a person places a thermometer under their arm for ten minutes. If the temperature observed is less than 97.8, this may be an indication of poor functioning of the thyroid hormone. Subsequent administration of small doses of supplemental thyroid glandular substance, which includes the active thyroid hormone, should normalize this temperature and alleviate many, if not all of the symptoms experienced.

When is adrenal gland supplementation necessary? The adrenal gland may become hyperactive as a result of under-active thyroid gland hormone. After a certain time of hyperactivity, the adrenal may not be able to keep up with the demands and become hypoactive itself. Therefore, we may then find that when a person first becomes hypothyroid, the adrenals, in response to the hypothyroid state, increase the production of adrenal cortical hormones in an attempt to produce a normal physiologic state and normal metabolic condition of the body. The adrenal gland is only able to do this for a limited length of time. After that length of time, neither the adrenal nor the thyroid is stimulating the body's metabolism at the correct rate. Therefore, as initial therapy, it is sometimes beneficial to add both thyroid and adrenal supplements. We often find that thyroid supplementation alone will result in the adrenal gland being able to take a holiday from overproduction of its hormones. This will subsequently leave the adrenal gland in a condition in which it can function normally.

If the thyroid functioning is low, how can it be ameliorated? There are both natural treatments and medical treatments for hypothyroidism. Several medicines in which man-made substances act to replace thyroid hormone exist. Synthroid is perhaps the most famous of these thyroid drugs. The thyroid drugs act to stimulate the body by their direct thyroid-like chemical activity, which affects the cells and tissues of the body.

Conventional medical doctors believe that these drugs are superior to glandular and natural thyroid products because they are more consistent in dose strength and activity as compared to the natural products. Alternative medical practitioners believe that it is often wiser to initiate needed thyroid supplementation with natural glandular substances that contain active glandular hormones.

Many of the adrenal and thyroid glandular products available do not contain the active hormonal ingredients due to FDA rules. In many of these supplements the actual active hormonal ingredients have been removed. It may very well be that supplementation with the rest of the gland after the active hormonal substance has been removed will nurture the thyroid and adrenal, thus improving the function and output of hormone. However, it is more likely that the entire glandular substance, including the active hormone, will yield a more beneficial result. Therefore, the use of natural thyroid or natural adrenal gland substance with the active hormone still intact is usually the best initial therapy.

This must, of course, be undertaken only upon a licensed physician's prescription. A physician can prescribe, after a careful analysis of a case, a natural glandular thyroid, which has active thyroid hormones present within the supplement. These will act to replace the missing or hypofunctioning thyroid hormone that the body lacks and very often leads to dramatic improvements in all kinds of energy and metabolism-related symptoms.

Physical energy, mental energy, stamina and vitality can be dramatically improved by natural thyroid supplementation. We often find that other seemingly unrelated medical conditions improve. Such conditions include a poorly functioning heart, elevated cholesterol, puffiness and fluid retention. These problems can be reduced by the beneficial metabolic effects of thyroid supplementation.

DHEA & Pregnenolone

DHEA and pregnenolone are very important elements in the anti-aging strategy. Pregnenolone and DHEA levels often fall as we age. As a matter of fact, DHEA levels gradually decline throughout adulthood and parallel the continuous aging process. They are "precursor" hormones and are found early in the production chain of most of the other important steroid hormones such as testosterone, estrogen, and progesterone.

Decreased production of pregnenolone and DHEA by the body may lead to diminished production of testosterone, estrogen, and progesterone. This will result in many of the signs and symptoms of aging such as deterioration of various body tissues and organs. Aging skin, sagging of the body, bone demineralization, and hair loss are but a few of the many deteriorations due to the loss of the body's youthful hormonal balance. This same loss of youthful hormones is also a causative factor in the lack of sexual desire, diminished sexual functioning, generalized weakness and lethargy, worsened mental functioning and many of the other common results of the ongoing aging process.

DHEA or "Dehydroepiandrosterone" is a hormone discovered fairly recently by scientists. This hormone has been found to have very important and perhaps even crucial effects, both for human health and development and in the aging process.

DHEA is produced by the adrenal gland and is a very prominent steroid hormone found naturally in the systems of young adults. Because DHEA can give rise to many other kinds of steroidal hormones, perhaps as many as ten or more, it is sometimes known as the "Mother Steroid."

DHEA is a precursor hormone to many of the other essential hormones produced by the body. By the term "precursor hormone," we mean a hormone that occurs very early in the chain of hormone production. Hormones are commonly produced in the body by starting with one form of the hormone and then modifying it to another form. A continued chain of modifications thus produces all the "downstream" hormones formed from such precursor hormones as DHEA and pregnenolone. For example, testosterone and estrogen are commonly produced after DHEA has been modified several times.

Since DHEA is found early in this chain of hormonal transformations, it is thought by many scientists to act as both a raw ingredient for the production of later hormones and to act as a buffer hormone. By "buffer hormone" we mean that DHEA can be converted into many other kinds of hormones and may act to follow the body to produce whatever other downstream hormones are most needed at the time.

Lack of adequate DHEA stores may cause the body to use different means of producing the final hormone it needs. This may cause the body to produce sub-optimal amounts of the final hormone. DHEA also seems to have direct effects on the body in its own right. These effects tend to be very potent in producing and maintaining a youthful state of the body. Although these effects are only now being slowly elucidated, the recent discovery that our cells contain specific receptors for DHEA lend credence to the idea that DHEA may have significant direct hormonal effects.

DHEA and its metabolites are very important in maintaining youthfulness and longevity. One benefit of DHEA is that it does not exhibit the same profound effects as other steroid hormones such as testosterone, estrogen, and progesterone. It is primarily used by the body to produce whatever amounts of downstream hormones are needed.

Although the majority of DHEA's effects on the body act as a substrate for the production of more complex downstream hormones, DHEA does have some direct effects. DHEA tends to de-excite the cells of the body. DHEA may have a modulating and de-exciting response to hormones, nucleic acids and other substances that may cause the cells to grow abnormally or to deteriorate with aging. It also appears to many to have "androgenic" or stimulating effects similar to testosterone. It seems likely that these apparent stimulatory effects are actually the result of increased "downstream" production of testosterone that often occurs when DHEA supplementation is given, rather than being direct effects of the DHEA itself. The combined de-excitation effects plus the androgenic effects are prime reasons that DHEA is an effective anti-aging therapy.

Growth hormone and DHEA, along with melatonin, are the two principal hormones produced by the body that decline in a direct relationship with the advancement of the aging process in humans. In other words, melatonin and DHEA production by the body are found to be at their peak in adolescence and early adulthood when humans are at their peak health and performance. Both of these hormones then begin a gradual decline until they are produced in only very small quantities, if at all, by the seventh decade of life. Therefore, it seems reasonable to assume that DHEA and melatonin may hold critical roles in the onset and continuation of the aging process in humans.

How can we determine whether individual levels of DHEA production are still optimal, as in our youth? One way is to simply consider our chronological age. After the second or third decade of life, DHEA production almost always falls steadily as each further decade ensues. In other words, those with normal, high levels of DHEA in their twenties and early thirties will usually have a decreased amount in their fourth decade of life and an even greater decrease by the fifth decade of life. Subsequently, there is usually a more substantial decrease by the sixth and following decades of life. It is very common for DHEA production to have fallen to nearly zero by the seventh decade of life.

Individuals often experiment by trying various levels of DHEA supplementation if they are past the age of forty. DHEA can be taken in doses ranging from 5 mg per day to 100 mg per day. Common replacement levels for older adults have been noted to be in the range of 25 to 50 mg for men and 15 to 30 mg for women. Depending on age, it may be beneficial to begin DHEA supplementation at the lower end of these ranges and then gradually increase the dose until beneficial effects are noticed.

If the increased dose causes any adverse effects such as nervousness, feeling overly hot, oily skin, or abnormal hair growth, then the dosage should be decreased into the optimal range. By trying the DHEA supplements, starting with a low dose and gradually increasing, one can determine whether they feel more energetic, more invigorated, or whether there are any noticeable negative side effects.

This is one of the simplest ways to determine whether DHEA supplementation is beneficial and what the appropriate dose should be. A more scientific way of coming to the same conclusion is a DHEA blood level test. Blood can be withdrawn from the body and tested for DHEA levels. These DHEA levels can then be compared to norms for various ages of life. It may be most beneficial, from a life extension perspective, to replace DHEA to a level found in young adulthood or early middle age adulthood. The level of DHEA found in young adults may or may not be tolerated by the aging body. However, replacing DHEA to the maximum point that the body can tolerate, up to the youthful levels, is thought by anti-aging scientists to be a very potent life extension therapy.

Resources of DHEA include man-made natural equivalents of the hormone and products produced from all natural sources such as plants. At this time the man-made natural equivalents seem to be the most potent and active forms available. The DHEA found in certain natural food substances, such as the Mexican yam, may not be as effective as the man-made natural equivalent DHEA. This is because the amount of DHEA present in natural food sources is minimal and the biological effectiveness in humans of the naturally occurring DHEA hormone-like substances is not entirely clear.

Pregnenolone is a crucial precursor hormone, essential to the body's production of other hormones. There are several places of hormone production in the body and pathways of formation for many of the major hormones. Pregnenolone is one of the earliest hormones produced in a series of chemical production steps in which other hormones such as testosterone and estrogen are made. Because pregnenolone is made into many other important hormones, its presence in the body has a great influence on the production and balancing of testosterone, estrogen and progesterone.

Potential beneficial effects of pregnenolone supplementation include memory improvement and increased mental functioning, enhanced thyroid function, a better "sense of well-being" and increased production of other hormones. Pregnenolone is also thought to assist the body in balancing the other major hormones. This balancing refers to the body maintaining the proper ratios of each hormone in relationship to the other hormones.

The supplementation of pregnenolone and DHEA is generally regarded in alternative medicine as a safe way to balance and stimulate the proper production by the body of other important hormones. In clinical practice, I tend to use pregnenolone most often in three instances. I use it to balance and stimulate the production of the other natural steroid hormones, to assist in hormone replacement for women after menopause, and in both sexes when mental deterioration is occurring. Pregnenolone is truly a very useful and gentle-acting hormone that will certainly become more widely appreciated as time progresses.

Telomerase therapy may allow us to re-set the genetic programs of the cells that produce the hormones of youth. Why do the hormone-producing cells curtail their production of the vital messengers of youthful physiology? Perhaps because they themselves age. Telomerase may allow us to retain the hormone-producing cells in a more youthful condition, thus encouraging those cells to continue hormone production far later in life then they do currently. This type of genetic re-engineering of the hormone-producing cells may well serve as a powerful way to get to the root of the problem of hormonal deficiency as we age.

Case History:

Normally, I recommend that anti-aging strategies be initiated in the third decade of life. Recently, however, I began anti-aging therapy for a forty-year old man. He had noticed some loss in vigor, energy, drive, and the ability to recover from heavy physical exertion. Weight lifting seemed to yield few positive results, but instead made him feel achy and tired. Lab testing revealed normal hGH for his age with somewhat less-than-normal testosterone for his age. Utilizing the theory that normal hGH values for those forty and over are likely physiologically sub-optimal, I began him on a program of oral hGH-boosting supplements tied to his exercise regimen. Within the first few days of therapy, he began responding in a positive manner. Strength, energy, and muscle gain began immediately. His energy, vigor and drive substantially increased. There are indications clinically that both his hGH and testosterone levels have increased, but remain in a safe range.

How to Increase Your hGH Levels

Individuals under the age of sixty can often increase their hGH levels without the use of expensive hGH injections. Many oral supplements contain natural substances that I believe can stimulate our own, natural, production of the hormone. On your own, you may wish to try this simple regimen, and then judge the effects for yourself. Consider trying powdered glutamine, two grams in water at bedtime on an empty stomach, every other night, after vigorously exercising earlier that day. This simple, inexpensive regimen can often boost the hGH levels naturally and safely.

Chapter Six
Sex Hormone Replacement Therapy

Replacement of falling levels of sex hormones in aging humans promises dramatic anti-aging effects. Both males and females commonly exhibit multiple signs and symptoms directly attributable to the loss of these essential hormones of youth. Recognized for decades as a prime reason for many of the ills and complaints of aging, the medical profession has a long history of providing sex hormone replacement therapy for older adults. Testosterone by injection has long been given to older men suffering from lack of sexual desire or inability to function normally. These symptoms of decreased libido and impotence may be improved in many men by the administration of supplemental testosterone or androstenedione. Additionally, the general vigor, strength, energy and lifespan may potentially be enhanced by testosterone administration.

Women offer an even more dramatic case for the benefits of sex hormone replacement therapy. While not all men go through "male menopause," all women do indeed experience the loss of female hormone levels in menopause. The attendant hot flashes, vaginal dryness, urinary irritation, skin deterioration, and bone demineralization are experienced to varying degrees by all post-menopausal women. Replacement of estrogen and progesterone has long been recognized as important to relieve troubling symptoms and actual physical deterioration in women after menopause.

Related hormonal and genetic (telomerase) therapies will be discussed in this chapter, including thyroid hormone, cortisol, prolactin, and vasopressin. Among these, thyroid hormone replacement is by far the most commonly necessary and most well known therapy, both in anti-aging practice and general medicine. All of these hormones work together, as well as with human growth hormone and DHEA. Dysfunction or lack of any of these hormones can result in problems specific to that individual hormone as well as causing dysfunction of the other hormones. The human body is a complex set of physiologic systems maintained and regulated by these hormones. A very important cause of the aging process in humans is undoubtedly the diminished function of these essential hormones of youth.

Sex Hormone Replacement In Males

Men commonly go through a period of dramatic change in sex hormone function similar to menopause in females. This drop in the amount or activity of male sexual hormones (androgens) is referred to as the andropause. Andropause has been described as "an indefinite syndrome composed of several constellations of physical, sexual, and emotional symptoms brought about by a complex interaction of hormonal, psychological, situational, and physical factors."(1) The andropause is therefore one of seemingly endless duration once it has begun. In this regard, the initial symptomatology of the drop in male hormone activity may seem to lessen with time, but the major negative effects such as decreased sexual ability and lack of energy remain to continually detract from the quality and length of life.

In addition to decreased potency and lessened libido, andropause can result in tiredness and easy fatigability. Men may also notice that their physique become less masculine in appearance. This may at times progress to the point of gynecomastia, an enlargement in the male breast to the extent that it resembles a woman's breast.

Several of the unwanted physical effects common in post-menopausal women also occur as prominent components of post-andropause males. Chief among these are osteoporosis, muscle atrophy, and skin deterioration. Osteoporosis is the gradual loss of bone tissue and strength that is very common in the elderly of both sexes. It progresses as the mineralization of the bone structures gradually decreases due to lack of hormonal stimulation after andropause. Although formerly not a well-known problem in elderly men, osteoporosis is steadily becoming recognized by the medical community as a significant factor in the health problems of older males. Just as it occurs in women, the gradual deterioration, demineralization, and loss of bone strength lead to increasing likelihood of fractures.

Measures of bone density in older men reveal steady declines with progressing age. As bone becomes less dense, its strength and rigidity markedly diminish. The resultant fragility of the skeleton makes traumas, such as falls, more likely to result in a fracture. It has been demonstrated that the incidence of fracture risk increases dramatically after andropause, doubling every half-decade. One of the most dangerous and troubling of fractures, fracture of the vertebrae of the mid-spinal column (thoracic vertebrae) is more common in men than women.

The true scientific reason for osteoporosis in men is not entirely clear. However, it appears that male hormonal functions and, in particular testosterone levels, is the dominant reason for male osteoporosis. The cells, which form and maintain bone tissue, the osteoblasts, are known to possess receptors for male hormones. These and other scientific evidence suggest that they are controlled, in at least some measure, by testosterone levels in the blood.

Further evidence for this is derived from young male patients suffering from a disease known as hypogonadism. Hypogonadism occurs when the male testes are unable to produce normal amounts of testosterone due to an abnormality in the brain's stimulatory messages to the testicles. This disease is associated with osteoporosis just as is andropause. Castration in young males also results in osteoporosis. Similarly, other disease states based on abnormal hormone levels and decreased testosterone blood levels, such as hyperprolactinemia, are associated with bone demineralization.

As can be readily seen, there appears to be a strong correlation between testosterone levels in the blood and bone strength and density. This observation leads us to imagine that replacement of testosterone in those diseases in which it is deficient will cause reversal of the osteoporosis. Indeed, this appears to be true. Scientific studies replacing male hormone in young men suffering from hormone deficits reveal that bone mineralization and bone strength improve markedly and osteoporosis is curtailed after treatment. In such male hormone deficiency diseases as hypothalamic hypogonadism, Kleinfelters Syndrome, hyperprolactinemia, castration, and anorexia nervosa, testosterone therapy reduces osteoporosis.

In elderly men suffering from osteoporosis, we find that the relationship between testosterone levels and bone mineralization also seems to exist. A significant percentage of men suffering spinal column vertebral fractures are deficient in male hormone. Likewise, bone mineral density studies involving the spine, fingers, and leg bones show the correlation of bone density to testosterone levels. Elderly male nursing home wards, who suffer bone fractures from light trauma much more readily than do young men, have significantly reduced blood testosterone levels.

Declines in the total amount of testosterone produced in the male as well as testosterone's ability to function normally appear to be the underlying reason for andropause. What causes this disruption in male sex hormone functions? Scientists have put forth evidence supporting several potential reasons.

The testicles themselves, (known to be the primary glands for testosterone production in males), may decrease in size and ability to produce sex hormones as men age. Another possibility is that chronic illnesses, which are more common as we age, cause diminished testosterone production by the testes. Perhaps the most intriguing theory involves the areas of the brain, which produce hormones that regulate and stimulate testosterone production. Sex hormone production by the testicles may deteriorate as these hormonal control centers of the brain become dysfunctional.

Areas of the brain known as the hypothalamus and pituitary produce hormones, which control the testicular production of testosterone in men. They manufacture hormones such as luteinizing hormone (LH) which direct the testes in their production of testosterone. LH is released into the bloodstream periodically as events called secretory bursts. Scientists can measure several different aspects of the secretory bursts, including the total amount release per burst, the time duration of each burst, and the amplitude of each burst. Studies indicate that decreases in one specific aspect of LH release from the brain, the secretory burst amplitude, result in diminished testosterone production by the testicles.

Likewise, diminished blood serum testosterone levels seem to lead to a greater frequency and length of LH release with each secretory burst. An alternative theory is that the testicles become less responsive to stimulation by LH. An overview of recent scientific studies points to the possibility that changes in the manner in which LH is secreted, or in the ability of the testes to respond to the LH, serve as the most significant bases for decreased testosterone production in older men.

Replacement therapy involves the physician-ordered administration of supplemental testosterone. Testosterone can be administered orally as pills, or by injection. Replacement therapy has been shown to benefit post-andropause men in many ways, including improving muscle strength, increasing bone mass, and improving mood and mental functioning, although side effects also occur. The beneficial effects of younger appearing skin, a more muscular physique, improved mental alertness and increased sex drive are the most striking benefits of testosterone hormone replacement therapy.

Side effects from testosterone replacement therapy include mild weight gain, sometimes to a noticeable extent in the breast regions. In some instances, apparently less frequent in older men, increased skin oiliness, acne and hair thinning may occur. Theoretically, the prostate gland will be stimulated just as are many of the important tissues of the body. This may lead to an increased chance of enlarged prostate (benign prostatic hypertrophy) or an increased incidence of malignant disease of the prostate gland, although there is at present no compelling evidence that testosterone replacement therapy in the elderly causes progression of preclinical prostate malignancy to clinical cancer. Neither is there current conclusive evidence of aggravation or causation of prostate hyperplasia (enlargement) in elderly men treated with androgen replacement therapy. It is recommended that a physician monitor the prostate by clinical examination and laboratory tests during testosterone hormone replacement therapy for any adverse changes.

The blood fats and ability of the body to metabolize carbohydrates may also be affected by testosterone administration. High density lipoprotein (HDL) is a beneficial form of blood cholesterol, which seems to help cleanse the arteries of atherosclerotic plaque. HDL appears to vary inversely with testosterone levels in young adults. Young men have higher testosterone levels and lower HDL levels. Young women have lower testosterone levels than young men, and also have higher HDL levels. These females also suffer heart disease much less frequently than do their male counterparts.

It isn't until later in life, when estrogen levels drop in women that heart disease incidence increases. The lower HDL in men may, therefore, be due to the males' higher testosterone levels, lower estrogen levels, or both. In young males suffering from hypogonadism, testosterone administration has led to increased incidence of sleep apnea. This effect does not appear to be prominent among those elderly to whom it is administered but may cause increased apnea in those who are obese, suffer from emphysema, or are heavy smokers.

Newer kinds of more "natural" testosterone administration, particularly when combined with genetic anti-aging threapies such as telomerase, may have the potential of far fewer negative side-effects, and may actually cause some very positive beneficial effects. Some physicians are reporting that the administration of these newer testosterone replacement and enhancement techniques actually lowers the total cholesterol and may lead to less, not more, atherosclerosis over time.

Sex Hormone Replacement in Females

Menopause serves as a dramatic and universal indication of sex hormone changes in older women. All women eventually undergo this cessation of menstrual cycles. The average age of menopause is fifty. Menopause is caused by the loss of youthful ovarian reproductive function (follicular development), which results in hormonal changes from the former, youthful, reproductive state. Estrogen hormone levels become markedly decreased, while the hormone, which vainly attempts to increase estrogen production, FSH, rises and serves as a clinical marker of the menopausal state.

Declining estrogen levels directly result in the cessation of menstrual periods, and the attendant symptoms well known to women who have suffered through menopause. Problems with "hot flashes," known in medical circles as vasomotor instability, dryness and atrophy of the female reproductive tract, osteoporotic bone loss, psychological depression, insomnia, and nervousness may all contribute to the constellation of symptoms and physical deterioration known as menopause.

Hot flashes are experienced by three-quarters of women as they undergo menopause. They are caused by short-term dilations of the small blood vessels and are perceived as a sensation of pressure in the head, followed by a feeling of heat and attendant sweating. These symptoms occur most commonly in the area of the head, neck, upper chest and back. Other prominent symptoms such as heart palpitations, faintness, dizziness, and fatigue are common. Individual women may experience varying degrees of symptomatolgy. The number of different types of symptoms as well as their frequency and duration may be drastically different in each woman.

What is the actual cause of these strange and troubling complaints? The current concensus within the scientific community is that diminished estrogen levels periodically cause a sudden and short-lived downward re-setting of the body's main thermostat, found in an area of the brain known as the hypothalamus. This then sets off a series of physiologic events in which the body attempts to dissipate heat by dilating blood vessels in the head, neck, and upper torso. This series of events then leads to the feelings of heat and sweatiness as the skin is involved in the process. Dizziness and faintness may result from dilation of the cranial vessels.

Estrogen replacement therapy employs oral, injectable, or dermal patch medications prescribed by a medical doctor. Additional sources for estrogen therapy may be found in natural plant sources. Plant-derived estrogen from the Mexican yam is currently available in health food stores and from some alternative healthcare practitioners.

Hormone replacement therapy with estrogen is well known to relieve the troubling symptoms of menopause. Additionally, physical deterioration from lack of estrogen can also be alleviated. Reproductive tract atrophy and deterioration is greatly improved with systemic or topical estrogen. The bone loss that results in osteoporosis and can result in bone fractures is markedly diminished by estrogen supplementation.

Another potentially dramatic health improvement from estrogen treatment is the prevention of cardiovascular disease. Estrogen is thought to raise the good cholesterol known as HDL, which then helps to clean the artery walls and decrease the likelihood of the development of atherosclerosis, heart attacks, and strokes. Perhaps the most important reason for the drastically lower incidence of heart disease in pre-menopausal women is due to this beneficial effect of estrogen. Replacement of the body's natural estrogen lost during menopause may then continue the cardiovascular protection that women enjoy from estrogen prior to menopause. Reports of drastic decreases in heart disease risks have been reported with post-menopause estrogen therapy. Epidemiologists have estimated a forty to fifty percent reduction in heart disease risk in these individuals.

Osteoporosis is one of the most medically significant problems of aging in women. Deterioration of the skeletal bones, demineralization and loss of strength and structural rigidity, leave older women at risk for fractures. These fractures can occur as a result of even fairly minor trauma. As a result of bone fractures, the elderly often suffer immobility or lack of physical capacity to perform the activities of daily living such as the inability of ambulates in the home, shopping for food and other necessities, and performing rudimentary activities associated with survival. These can be devastating to the patient's health and well being. Acute life-threatening medical emergencies such as pulmonary embolism can also result from traumatic fracture.

How can osteoporosis in women be avoided or minimized? Estrogen replacement plays a key role. Maintenance of adequate blood estrogen levels allows the bone to keep its mineral content. Calcium is therefore retained within the bone keeping the skeleton strong and resistant to fracture. Supplementation with calcium and vitamin D are also important to bone health and the minimization of osteoporosis.

Bone loss begins after age forty in men and at menopause in women and normally continues into senescence. To some extent, high dietary intakes of calcium along with adequate vitamin D can interfere with this bone demineralization. However, the addition of estrogen in post-menopausal females markedly diminishes the daily requirement of calcium needed.

Women on estrogen need to consume eight hundred or more milligrams per day of calcium, whereas women without estrogen replacement therapy are recommended to consume at least 1000 to 1500 mg per day. Other factors influencing osteoporosis include dietary habits, exercise habits, cigarette smoking and alcohol consumption. High fiber intake, although beneficial in many ways, can worsen osteoporosis if the fiber intake is excessive. Excessive dietary intakes of protein and phosphorous can also harm bone mineralization.

In an attempt to minimize the risk of increased incidence of endometrial malignancy, physicians are now offering therapy that includes both estrogen and progesterone. A common prescription includes estrogen such as Premarin for the first twenty-five days of each month along with a progetin such as Provera on days fifteen through twenty-five or days eleven through twenty-five of the same month.

A more recent therapeutic regimen involves the continuous use of a medication that combines estrogen with low dose progesterone. This is easier to take and results in less withdrawal bleeding. Menstruation-like withdrawal bleeding is probably the most negative aspect to women on oral estrogen and progesterone treatment.

Potential risks of balanced treatment seem to be continuations of the risks associated with the body's premenopausal production of estrogen. This hormone acts as a stimulant to many tissues, and the body's own naturally produced hormone may aid in the growth of some cancers such as breast malignancy.

A possible slight increased incidence of breast cancer with medically prescribed supplemental estrogen is greatly debated in the medical world. Years of medical studies have as yet failed to provide conclusive proof of this potential risk. However, the risk of increased incidence of endometrial cancer in woman who have not had their uterus removed by hysterectomy is well established. There is an approximately six-fold greater chance of this malignancy in women who are given estrogen supplementation without additional progesterone hormone. This dual hormone therapy is therefore commonly prescribed in women who have not undergone hysterectomy.

Additional potential problems associated with estrogen replacement therapy include an increased incidence of a certain blood fat known as triglyceride and an increase in gallstone disease. Biomedical scientists are currently exploring the possibility that estrogen replacement therapy may also increase blood pressure and the clotting ability of blood.

Related Hormonal Therapies

Thyroid disease takes on greater consequence, and becomes more difficult and complicated to detect as people age. Elderly persons are known to have a higher incidence of goiter and thyroid nodules. The symptoms of thyroid disease may be subtle and difficult to ascertain in older individuals. It is important to detect low thyroid (hypothyroidism) or elevated thyroid hormone output (hyperthyroidism) in older persons as these conditions can have even more dire consequences in the elderly than they do in younger patients.

Either hyperthyroidism or hypothyroidism can cause congestive heart failure, which is a life-threatening condition caused by a weak heart. In like manner, cardiac rhythm disturbances such as atrial fibrillation can result from hyperthyroidism. Atrial fibrillation can lead to blood clots, known as emboli, being promulgated in the blood vessels. Stroke and other serious conditions can result from this phenomenon.

Interpretation of thyroid lab tests is more complicated in the elderly. Physicians commonly test several sub-types of thyroid hormone including T3, T4 and T7. The standard reference value ranges are applicable to the elderly for these tests, but another important test, that of thyroid stimulating hormone (TSH), is more difficult to interpret when compared to standard reference values. This is particularly important as TSH is considered the single most important test to detect thyroid disease in elderly people.

The normal, healthy elderly person will commonly have a low value of TSH even though thyroid disease is not present. This makes the physician's job of interpretation of the TSH more difficult and makes the other tests such as T3, T4, and T7 along with symptomatology all the more important to come to a correct diagnosis. Hypothyroidism, which in younger people results in an elevated TSH, may result in a normal TSH in the elderly due to their blunted TSH secretion response.

Thyroid treatment is also trickier in older patients. Thyroid replacement therapy needs to be instituted more slowly and with greater caution than is normally required. Underlying cardiovascular disease can be made symptomatic by a rapid initiation of thyroid replacement therapy. Angina pectoris or other more serious sequelae may result.

Prolactin levels increase after age fifty in men and age eighty in women. This hormone, secreted by the pituitary, appears to be controlled to at least some extent by the dopaminergic system of the central nervous system. Deterioration of the dopaminergic control system in aging may potentially be lessened or delayed by the MAO-B inhibitor Deprenyl. Deprenyl is the topic of discussion in a separate chapter of this book.

What is the significance of increased prolactin production? In older men, elevated prolactin has been linked to the diminished sex drive and interest in sex so common in older men. Cortisol and vasopressin levels are less influenced by the aging process.

The cortisol axis seems to be unperturbed by senescence except for the hormone dehydroepiandosterone (DHEA). This hormone is known to decrease with advancing age and is discussed in detail in another chapter. Vasopressin is a hormone that helps the body regulate fluid and electrolytes (salts). It is well known that older individuals have a greater difficulty maintaining this balance than do younger people and are more prone to fluid retention, edema, and swelling.

However, scientific studies show that total vasopressin levels do not change significantly with age. Some studies show that under certain physiologic stresses, vasopressin responses of elderly subjects are different from those of younger subjects. At this time, it appears that the primary reason for the diminished fluid and electrolyte handling capability of the elderly lies in a decreased kidney function, rather than vasopressin abnormalities.

Genetically, our cells may be able to retain their ability to manufacture sex hormones longer, and respond to the presence of sex hormones in a youthful manner, if telomerase therapy is successfully employed. Restoring the genetic anti-aging effects of telomerase within our cells may make the replacement of the sex hormones unecessary, or at least delay the need into later years. It seems plausible that allowing our own cells to make natural sex hormones longer is the wisest path to retaining the most youthful form and function of our body.

Case History:

It is well known to all of us that men often respond in a dramatic way when we increase their testosterone level. Less known is that women respond even more so. A fifty-year old woman patient of mine complained of poor energy, a negative sense of well-being and of having no sex drive. Often, a lack of libido is indicative of other medically significant problems.

I began by measuring her female hormone levels. Testing showed low estrogen and progrsterone as well as DHEA, pregnenolone and also, significantly, testosterone. Replacement of pregnenolone, DHEA, and progesterone led to a greater abundance of estrogen in her system. However, a small dose of natural testosterone was required to really bring an end to all her symptoms, including the libido problem. She was surprised to learn that a small amount of what is usually thought of as the "male" hormone was an essential part of making her feel once again, as she told me, "like a woman."

How to Replenish Sex Hormones

Perhaps the wisest place to start is by trying to stimulate our bodies to produce more of our own, natural, sex hormones. An interesting way to do this is by combining exercise and growth hormone stimulation to produce a secondary cascade of sex hormone production by our body. The noticeable effects of increased sex hormone production such as increased libido and improved sexual performance are often readily and immediately apparent.

This regimen also often yields surprisingly dramatic ancillary benefits such as the lessening of body fat and improvement in heart function. Some also notice less wrinkles of the skin and sagging of the body. How do we attempt to achieve all these goals? By exercising thirty minutes every other day (half aerobics and half moderately intense weight training) and by simultaneously taking a special supplement regimen. Creatine monohydrate is used one tablespoonful or more (ask your physician for specific recommendations for you as an individual) before and another dose after each workout.

Additionally, at bedtime of each day the workout was performed, we utilize a good growth hormone stimulator such as "GH Boost" plus niacin 250-500mg time released. The GH Boost and niacin are taken at bedtime on an empty stomach on the nights the workout was done earlier in the day. This often results in a stimulation of growth hormone and also a secondary cascade of greater sex hormone production, especially the "androgenic" hormones such as testosterone, DHEA and androstenendione. In some patients we further augment the androgenic response by administering supplemental androstenedione. I advise caution and to use these ideas in a careful and judicious manner, seeking specific advice from your personal physician.

References

1. Henker, F. D., “Sexual, Psychic, And Physical Complaints In Fifty Middle-Aged Men." *Psychomatics* 18 (1977): 23.

2. Veldhuis, J. D., Urban, R. J., Lizarralde, G., Johnson, M. L. and Iranmanesh, A. “Attenuation Of Luteinizing Secretory Burst Amplitude As Approximate Basis For The Hypoandrogenism Of Healthy Aging In Men.” *Journal of Clinical Endocrinology and Metabolism* 75 (1992): 1, 52-58.

3. Tenover, J. S., Matsumoto, A. M., Plymate, S. R., and Bremner, W. J., “The Effects Of Aging In Normal Men On Bioavailable Testosterone And Luteinizing Hormone Secretion: Response To Clomiphene Citrate," *Journal of Clinical Endocrinology And Metabolism* 65 (1987): 4, 1118-25.

4. Riggs, B. L., Wahner, H. W., Dunn, W. L., Mazess, R. B., Offord, K. P., Melton, L. J. “Differential Changes In Bone Mineral Density Of The Appendicular And Axial Skeleton With Aging”, *Journal Of Clinical Investigation* 67 (1981): 328-335.

5. Bardin, C. W., Swerdloff, R. S., Santen, R. J. “Androgens: Risks And Benefits." *Journal of Clinical Endocrinology And Metabolism* 73 (1991): 4-7.

6. Lobo, R. A., "Estrogen And Cardiovascular Disease," Annual *New York Academy Of Science* 592 (1990): 286.

7. Weitzman, A., Weitzman, R., Hart, J. "The Correlation Of Increased Serum Prolactin Levels With Decreased Sexual Desire And Activity In Elderly Men." *Journal Of The American Geriatric Society* 31 (1983): 485.

SECTION FOUR - BODY AND MIND

Chapter Seven
SEXUAL FUNCTIONING AND SEXUAL ENERGY

Deterioration in sexual functioning and lessening of normal sex drive is a common occurrence during the aging process. Decreased "libido" or sex drive is often noted as people age. In addition, the actual physical sexual functioning of the sex organs deteriorates as humans age.

Scientists have discovered that, in both men and women, the production of sexual hormones gradually declines after the second decade of life. In males, testosterone hormone begins declining after the late teenage years. In women, the main female sex hormone, estrogen, begins declining in the third decade of life and falls off until a low level is reached. Menopause then ensues. It appears very likely that the declines in sexual hormones and other related hormones of the body play a very significant role in the diminishment of sexual functioning seen in older adults.

In addition to decreasing hormone levels, other functions, particularly those of neurotransmitters within the brain, decline or change with age. This may lead to decreased sexual function and physical deterioration of the sexual organs.

Some of the neurotransmitter chemicals used by the brain's nerve cells for communicating with each other are acetylcholine, dopamine, and norepinephrine. Fluctuating levels of these neurotransmitter chemicals with aging can cause the brain's functioning to diminish. In particular, the brain controls a significant part of the sex drive of individuals and also their sexual performance abilities. Not only do these neurotransmitters directly influence the sex drive and sexual functioning, but may at times directly or indirectly play a role in the control of the production of the major sex hormones such as testosterone and estrogen.

There are certain clinical indicators of the need for hormone replacement therapy with sex hormones. These include decreased sexual desire, decreased sexual functioning, and atrophy of the sex organs. In males, there may be a shrinking of the testes, a loss in muscle mass, and a decreased libido or sexual desire. Finally, there may be decreased erectile functioning of the penis.

Women may notice signs of the need for hormone replacement therapy with sex hormones when they experience decreased libido or desire for sex, the vaginal lining becomes thinner, weaker and poorly lubricated, and that the quality of resiliency of their skin begins to worsen.

Laboratory testing can perhaps be one of the best indicators of the need for replacement of sex hormones. Blood levels of testosterone, estrogen, progestins and prolactin can be measured and judged against normals for age. Likewise, salivary testing of these hormones is also available. These test results can serve as an indication for needed hormone replacement therapy to improve sexual functioning. The primary sex hormones that may benefit by hormone replacement therapy are testosterone, estrogen, progesterone, DHEA, and prolactin.

A powerful sex hormone replacement therapy for women is natural progesterone. Natural progesterone, especially when combined with pregnenolone, may help to replenish and balance all of the sex hormones found in females. This process occurs through two processes. First, the progesterone itself will have direct beneficial effects and may cause a more healthy balance to occur vis-a-vis estrogen when a "hyperestrogenism" condition exists. Secondly, the female body may be able to convert the progesterone into other needed steroid hormones such as estrogen and testosterone. Pregnenolone, if present in the supplement, may also aid the body in producing other hormones in which it is deficient. This conversion of progesterone and pregnenolone into deficient downstream hormones occurs because the progesterone and pregnenolone can serve as the raw materials for the production of the other hormones.

Natural progesterone cream is produced from the roots of wild yam plants. A two-ounce jar of Progest Cream will commonly contain approximately 900 mg of progesterone. Progesterone cream is massaged into the skin until it is absorbed. This allows for a slow absorption of the progesterone, which makes its effects much gentler and less shocking to the body.

It is useful to apply the progesterone to areas of the body that have softer, thinner skin and to rotate the areas to which the cream is applied. Natural progesterone has been found by chemists to be an exact duplicate of the progesterone that the body produces. In contrast, medical progesterone is not exactly identical to the progesterone the body produces. Medical progesterone mimics the effects of our natural progesterone. Due to the fact that it is a man-made and altered chemical, it has side effects different from those of natural progesterone.

In addition to natural progesterone cream, natural progesterone oils are available. Natural progesterone oil is a combination of natural plant progesterone in vitamin E oil. The mixture of vitamin E with progesterone causes the progesterone to act in a different manner than the progesterone cream.

The vitamin E and progesterone mixture is absorbed into the body's lymphatic system. This allows it to pass more readily through the blood stream and then to the various tissues of the body before being metabolized by the liver. It can be used under the tongue, massaged into the soles of the feet or used vaginally. Either or both the progesterone oil or progesterone cream can be used. A one-ounce bottle of progesterone oil contains a large amount of progesterone. This amount is approximately 3,000 mg. The concentration is commonly three mg per drop of progesterone oil.

Another beneficial agent for women to use along with natural progesterone are the natural estrogens. They can also be obtained from plant sources and are sold as a cream. The use of both natural progesterone and estrogen together can help balance the female hormonal system. They can enhance sexual drive and sexual functioning. Estrogen also directly helps to strengthen and repair the lining of the vagina.

Both men and woman can benefit from administration of testosterone or androstenedione. Testosterone is essential to the functioning of both sexes, although men require a larger amount. Androstenendione is the immediate hormonal precursor to testosterone in the body. Not only does testosterone cause the body to maintain many of its organs and tissues, but it also is essential for sexual libido in both men and women.

Following menopause in women, and commonly after age forty in men, testosterone levels drop. These individuals may benefit from testosterone supplementation. Common daily dosages range from 10 to 20 mg for women, and 50 to 400 mg for men. Testosterone levels can be measured by blood and saliva tests. These same blood or saliva tests can be used to determine the correct continuing doses of testosterone augmentation.

Another important hormone related to sexual functioning and sexual drive is DHEA. DHEA has been called the mother steroid. It gives rise to many different steroid hormones in humans. DHEA and the hormones produced from it are essential in maintaining proper sexual functioning and youthfulness of the tissues of the body. DHEA levels can be measured by blood or saliva tests. It can be fairly easily supplemented. If a multiple hormone replacement therapy for sexual functioning is employed, it should be noted that the various sexual hormones supplemented will all affect each other's levels. Therefore, their levels should be carefully monitored.

A new medication, sildenfil (Viagra), promises hope for male sufferers of impotence. The manufacturer, Pfizer Corporation, states that full normal sexual function can be restored to many impotent men. This medication works by blockage of an enzyme, which aggravates impotence. Sildenfil is to be taken by mouth approximately an hour before an erection is desired. The medication, combined with adequate sexual stimulation, may lead to normal sexual functioning in the otherwise impotent male.

The medical drug, Yohimbe hydrochloride, is used to treat impotence. It seems to work by changing the penile blood flow patterns and thus facilitates penile erection. Yohimbe has been successfully used to treat impotence in men whose sexual dysfunctions stemmed from psychogenic (mind related), vascular (blood vessel related), and diabetic origins. A positive side effect of the drug is that it elevates mood in some patients. Potential negative side effects include anxiety, nervousness, irritability, elevated blood pressure, increased heart rate, and tremor. Yohimbe hydrochloride is available under the brand names Dayto Himbin, Yocon, Yohimex, Aphrodyne, Prohim, and Yovital.

Besides hormone replacement therapy, there are two other kinds of biotherapies available to the individual interested in improved sexual function. These two are homeopathic remedies and herbal remedies. Herbal remedies can act as potent sexual enhancers. Certain herbs can increase the sex drive and improve sexual functioning. This occurs by improving the functioning of the mind and the functioning of the various sexual organs and tissues of the body involved in sexual activity.

There are perhaps two basic ways in which to use herbal supplements to increase sexual health and sexual functioning. The first way is to use single herbs on a short-term basis to temporarily increase sexual drive and functioning. Yohimbe is effective in this regard. It is a stimulant of male sexual functioning and has also been used in longer-term regimens. Other single herbs shown to enhance the glands of male sexual functioning include damiana, black walnut, sarsaparilla, hoshouwu, saw palmetto, Siberian ginseng, American ginseng and Suma.

Likewise, for female sexual functioning, Dong Quai may be useful. Dong Quai has been used for centuries as a natural herbal sexual rejuvenator for women. It has been used to improve libido and functioning of the female sexual organs, as well as calming the mind and providing an improved sense of well being.

Homeopathic remedies are natural substances prepared in such a way as to allow them to cause beneficial effects within the body. Carefully selecting the proper homeopathic remedy for an individual suffering from sexual dysfunction must be done by a well-trained physician.

The doctor will commonly consider at least two primary factors when evaluating the individual. First, and perhaps most importantly, the individual's mental and emotional state are analyzed. What kinds of psychological factors may be involved? How does their deepest mental and spiritual being affect the sexual functioning of their body? Secondly, the actual physical symptoms of the patient are important to analyze when prescribing the correct remedy. Many general physical states occurring as we age can indirectly affect sexual function. Other physical maladies directly affect it. All of these factors are carefully considered when selecting the appropriate homeopathic remedy.

Sexual Anti-Aging in the 21st Century

As we enter the new millenium, many aspects of the aging process are giving way to advancements in anti-aging medicine. Perhaps the therapies that restore or conserve healthy sexual functioning are among the most interesting to the majority of people seeking longevity enhancing treatments. It seems likely that many exciting new therapies will soon enter the arena of modalities available for physicians specializing in life extension medicine.

Paramount among these is telomerase. As newer and more varied types of telomerase therapy are developed over time, innovative ways to enhance sexual health will emerge. Applying these advances to the lives of their patients may well yield benefits beyond the mere mundane. A youthful sexual outlook and ability may stimulate both the psyche and the body as individuals retain a positive perspective on life!

How To: Viagra and Natural Enhancers

Medications and natural supplements that modulate the way our blood vessels expand and contract are at the forefront of sexual medicine today. Our vascular system must be able to respond to sexual stimulation by changing the blood flow to the sexual organs in order for us to have normal sexual function.

Viagra, a pharmaceutical drug, helps to restore the normal ability of the penile and vaginal blood vessels to respond to sexual stimulation. This results in better erections for men and more profound sexual responses for women. Yohimbe, the most famous of the natural sexual enhancers, may also use a similar mechanism of action for some of the effects it causes.

My experience with patients has taught me that people attain the best responses with these substances if they are used intermittently, rather than continually. Coordinating the timing of doses of Viagra or yohimbe to coincide with likely sexual activity yields the best results. Other stimulatory natural substances such as androstenedione also seem to work best when used intermittently for more than a few days at a time.

I have seen the best results when both men and women use smaller doses of androgenics such as testostereone or androstenendione for a few days at a time, correlating with time periods when sexual activity will be most likely.

Timing the use of agents that have a shorter duration of action is also important. Viagra and yohimbe may work best when taken within minutes or hours of sexual activity. Combining the longer-term and shorter-term actions of the various agents available seem to produce the most exciting effects for those with impaired sexual functioning.

Case History: Viagra

Most of the cases in this book are about people who have succeeded in their quest to improve their health or physical functioning. However, sometimes it is instructive to hear the other side...about individuals who have not done as well.

A few years ago one of my patients, a man in his seventies who suffered from many and varied health maladies, wished to try Viagra to improve his sexual functioning. He had major problems with achieving an erection. Unfortunately, his other severe health problems contraindicated many of the sexual enhancers.

Since his heart seemed to be in decent shape, we tried Viagra. He tried a Viagra one night, and then the next. Nothing. Thinking he may have been given the wrong medication he returned to the pharmacy and asked them, "Is this really Viagra?" It was. He simply had not responded. Why? It could have been many things. Perhaps his vascular system had lost the ability to respond to the drug. Or perhaps other factors such as nerve impairment, emotional or other physical factors were the real impediment to his sexual functioning. In any event, it goes to show that sexual response is a very complex and delicate mechanism.

On a more positive note, another man was able to achieve marked increase in sexual functioning by simply taking a small dose of androstenedione intermittently. Why take it intermittently instead of all the time? Because the body may become habituated to it. If this happens, a higher and higher dose will be required to produce an effect. As we progress up the dosage ladder, the chance of negative side-effects rises. With androstenendione, prostate problems could potentially develop, among others. It is important to use the sexual enhancers in a careful and thoughtful manner if one desires the best results and to minimize the likelihood of problems.

Chapter Eight
ANTI-AGING THERAPIES FOR THE SKIN

Cosmetic anti-aging medicine is primarily concerned with ways of ameliorating the changes wrought upon the skin and supporting structures as we age. It is well known that as humans age, the skin, especially the skin of the face and other exposed areas, become wrinkled, thickened, blotchy in coloration, and prone to the development of acne.

Along with the gradual deterioration of the skin itself, layers underlying the skin also deteriorate with age; these are known as the subcutaneous layers. The most significant subcutaneous layer, as far as total volume, is the subcutaneous adipose layer, which is composed of fat.

Aging changes affecting the adipose subcutaneous layer include larger volumes of deposits of fat, particularly in the areas of the abdomen and hips. Additional changes in the fat layer include lumpiness and the development of benign fatty growths known as lipomas. Altogether, the degeneration due to aging causes sagging and laxity of the skin and subcutaneous tissues, resulting in what is commonly known as middle-age bagginess. This bagginess primarily affects the face and facial structures, the upper arms, hips and buttocks.

These changes due to aging usually become noticeable by the middle of the third decade of life. The most prominent area in which these changes become noticeable is in the face. The face, being exposed to sunlight and the wrinkling effects of facial expressions, is perhaps the most sensitive area of the body to cosmetic aging changes.

The following is a list of commonly noticed aging changes for men and women ages thirty-five to forty-five. It is interesting to take an accounting of our own personal condition. Place a mark next to each of these changes that you may have noticed in yourself:

_____ Gradual hooding of the upper eyelids
_____ Fine wrinkling around the eyes
_____ Puffiness under the eyes
_____ Fine wrinkles forming at the corners of mount and lips
_____ The beginnings of vertical frown lines between the eyebrows
_____ Appearance of nasolabial folds or creases, beginning at the outer corner of the nose and extending to the sides of the mouth
_____ Muscle weakness and loose skin noticed in the abdominal area
_____ Localized fat accumulation, predominately in the hips, thighs, buttocks, and abdomen
_____ Loss of volume and tone in the breasts

Next, check the following findings that commonly occur in men and women from ages forty-five to fifty-five:

_____ Bags under the eyes
_____ Sagging of the upper eyelid
_____ Crows feet of deeper wrinkles around the eyes
_____ More extensive and deeper wrinkling of the forehead
_____ Double chin
_____ Sagging of the eyebrows
_____ Vertical cord like structures forming in the cord of the neck (turkey neck)
_____ Deepening folds in the nasolabial area with deeper wrinkling around the mouth
_____ Sagging of the soft tissues around the jaw line causing the development of jowls
_____ Minimal drooping of the tip of the nose
_____ Increased sagging and fullness in the lower abdomen and thighs
_____ Breast sagging with low nipple position

The skin is an organ of the body that is very susceptible to the aging process. One of the prime reasons for this is the skin's exposure to sunlight. There are many reasons the tissues and organs of the body undergo aging. One of the leading scientific theories of aging is the theory of cross-linking. Cross-linking is the chemical interaction of various components of the tissues that cause them to form abnormal links between each other. These links are unlike the normal structure that the tissue was designed to have. Sunlight is a prime causative factor of such cross-linking in tissues.

Cross-linking of the tissues of the skin is thought to be one of the main mechanisms for aging of the skin. As the skin ages, it becomes thinner, forms wrinkles, and develops blemishes. These blemishes can include darkly pigmented areas and a type of acne known as rosacea. Along with cross-linking due to sunlight, hormonal factors and other factors cause aging of the skin. Since our face, neck, arms and upper body are the most commonly exposed to sunlight, these are the areas that age more rapidly.

Particularly noticeable and problematic for women, this aging of the face, neck and upper body seems accelerated. This happens because women's skin is thinner and more delicate. The skin of the male is somewhat thicker, tougher and more resilient, but still shows dramatic signs of aging as the decades pass. There are several different anti-aging therapies that are available to those concerned with the aging of their skin and the cosmetic and health effects due to the aging process.

Retin A

Retin A is a prescription medicine form of Retinoic Acid. Retinoic Acid is a derivative of vitamin A. Retin A, therefore, is a close relative of vitamin A, which is known to be a strong antioxidant vitamin. Effects of vitamin A on the skin include increasing the normal, youthful collagen-type fibers deep within the dermal layers of the skin. It decreases wrinkling of the skin, particularly the fine wrinkles, and it clears some of the darkly pigmented areas of the effected skin.

Retin A comes in cream or lotion form and is available in several different strengths. Retin A is commonly applied once daily at bedtime. One of the common mistakes of using Retin A is applying it too heavily or using too strong of a dose. As an anti-aging treatment, lower doses of small amounts of the lower strength doses of Retin A massaged gently into the skin appear to be the wisest form of treatment. Using smaller amounts of lower potency Retin A causes the medicine to take longer to produce the beneficial effects. However, it is much safer when used this way and may give the cells of the skin more time to adjust to their new metabolic environment.

Drawbacks of using Retin A cream in a too high strength include, 1) burning and exfoliation of the skin, and 2) the skin can turn red, hot and feel as if it has been burned. The exfoliation occurs as the shocked cells of the skin cause the outer layers, known as the epidermis, to slough off. This can lead to a dramatic, ugly, and unhealthful appearance and condition of the skin for quite some time after application of high doses of Retin A.

It is therefore advisable for Retin A to be used very conservatively and to realize that it may take weeks or months for improvement to occur. This slow improvement may be safer and perhaps longer lasting. In order to maintain the improved and more youthful skin produced by Retin A, the medication must be used for an indefinite period of time.

Alpha Hydroxy Acids

Alpha Hydroxy Acids, also known as AHAs, are naturally occurring acids found in many plants and fruits. This group of natural plant acids has been employed for hundreds of years as a potent cosmetic medicinal to moisturize and rejuvenate the skin. Plants from which alpha hydroxy acids are most commonly taken include citrus, apples, grapes, and sugar cane.

Alpha hydroxy acids work by diminishing the glue-like substances that cause dead skin cells to adhere to the skin surface. As the AHAs loosen these dead cells from the skin, several beneficial things happen. The living layers of skin cells are able to "breathe" and regain a healthier, smoother, softer appearance. The pores of the skin are also able to drain better, which lessens the occurrence of acne. Finally, the remaining layers of living skin cells are better able to metabolize. They thereby function in a healthier way. This improved functioning may allow the skin to age at a slower rate. Cosmetic problems alleviated via these processes include dry skin, acne, weathered and sun damaged skin, fine wrinkles, and blotchy pigmentation.

Cosmetic benefits of alpha hydroxy acids are gained as the dead layers of skin cells are sloughed off. The remaining skin then has a much smoother texture and a more uniform color. The complexion appears rough and dull prior to treatment due to the accumulated build-up of dead skin cells. When AHAs are used repeatedly, they stimulate the growth of new skin cells. These new layers of skin cells are actually plumper, which produces a tightening of the skin and reduces wrinkles. The plumper cells also appear healthier and more youthful.

Unlike certain other skin rejuvenating products such as Retin-A, AHAs do not usually cause the skin to become overly sensitive to sunlight. When used correctly, AHAs seldom cause significant redness, irritation, or flaking. As the skin cells become healthier, they may actually become more resistant to irritants such as chemicals.

Many skincare products contain alpha hydroxy acids. These include creams, astringents, lotions, cleansers, gels and skin peels. Peels are skin treatments in which a solution containing concentrated AHA is applied to the skin. After a brief time has passed, the solution is washed away. This process results in sloughing of several of the upper most layers of skin. The sloughing process takes several days after the original treatment to complete. After the old layer has peeled off, the healthier, "living," underlayers are exposed. The newly exposed fresh skin has a healthier color and texture.

Peels can be superficial or deep. Deep peels cause a more sudden and profound exfoliation and may result in the patient missing several days of work or other normal activities. Deep peels may cause seven to eight days of active exfoliation. Superficial peels cause a milder flaking of the skin for one or two days and are better tolerated. Light superficial peels usually require several treatments to achieve the same total exfoliation as is derived from a deep peel. Results are commonly noticed within two or three weeks after treatment.

Areas of the body amenable to treatment include the face, arms, hands, neck and chest. As a matter of fact, nearly all of the skin surfaces of the body can be improved by the use of AHAs. Alpha hydroxy acid peels do not permanently lighten the skin. Therefore, they can be used in people of color including Asians, Africans and Hispanics.

"Light" chemical peels can also be performed using glycolic acid. Peels with glycolic acid are more superficial than are medium to deep peels, which can use either chemicals or plant-derived acids. Glycolic acid “light peels” cause a rejuvenative effect, which improves the cosmetic appearance of the skin. They improve the texture of the skin and seem to reduce fine wrinkling due to the removal of old layers of skin cells.

It commonly takes thirty to sixty minutes for this in-office procedure. Home treatment is also available using glycolic acid. The procedure is usually tolerated well. The person undergoing the peel may notice a brief period of tingling and slight redness of the skin following the procedure. Makeup can be worn immediately after the procedure and individuals can return to their normal daily activities, including work, immediately.

Medium to deep chemical peels can be performed with plant acids, phenol, or trichloroacetic acids. The rejuvenative effect of the chemical peels with trichloroacetic acid or phenol is caused by smoothing of wrinkles and the diminishment of blemishes on the skin. Unevenly pigmented or sun-damaged skin of the face is also partially rejuvenated as the chemical peels cause the outer layers of the skin to slough off.

This procedure is usually done in a physician's office. It takes two to three hours for a full facial procedure. The patient's recovery phase is marked by some swelling, itching, redness, and sensitivity to the sun in the areas of the skin treated. It usually takes five to ten days for the new skin to form after a medium depth peel. Following a deeper peel, new skin commonly forms one to three weeks after the procedure.

The redness noted after deep peels usually subsides in three to six months. It is recommended that makeup not be worn until a week after a medium depth peel and for two to three weeks following a deeper peel. People can usually return to work and other normal daily activities in five to ten days following a medium depth peel and two to three weeks following a deep chemical peel.

Dermabrasion is another medical procedure that can have a significant rejuvenative effect on the skin. This rejuvenative effect can enhance the cosmetic appearance of the skin through the diminishment of fine wrinkles of the skin. Fine wrinkles that can be especially benefited are those above the upper lip. Dermabrasion also improves the overall skin texture and diminishes acne and facial scarring.

Dermabrasion acts through the physical removal of the top layers of the skin by the use of a high-speed rotary wheel. The dermabrasion procedure commonly takes sixty to ninety minutes. A local anesthetic is commonly given, sometimes with sedation or general anesthesia. This procedure is usually done in a doctor's office as an outpatient or in an outpatient surgery center.

The recovery phase following dermabrasion involves temporary swelling, redness, burning and itching in the areas treated. The redness usually fades after several months. The treated areas of skin remain sensitive and become lighter in color where treated. Patients can wear makeup one to two weeks following the procedure. They can return to work and other normal daily activities one to two weeks following the procedure. It is recommended that patients who undergo dermabrasion avoid strenuous activity for a period of two to three weeks following the procedure.

Topical Vitamins

Topical vitamin formulations are currently a hot topic in cosmetic anti-aging medicine. Formulations in which vitamins and other natural rejuvenating substances are introduced to the skin include Cellex-C, C-esta, glycolic acid products, and Ethocin. The general principle behind their use is that many vitamins and natural substances can act as antioxidants and rejuvenators of the dermal (skin) tissues.

Vitamin C, bioflavonoids, vitamin E, and glycolic acids are some of the most commonly employed natural substances. Application of creams or lotions containing these anti-oxidant vitamins is surmised to allow the vitamins to slowly soak into the skin. The beneficial effects of the vitamins is thereby concentrated directly into the dermal tissues most needing their salutary actions.

Collagen Enhancement

Collagen products introduce additional protein into the dermal (skin) layers. Collagen is the naturally occurring protein which is the primary constituent of the under layer of the skin known as the "dermis." In contrast, the upper layer of the skin is known as the "epidermis." Collagen proteins form a network of fibers, which provide the framework upon which the skin cells and blood vessels that supply the skin live. Collagen acts as the supporting framework for the skin, just as our bony skeleton supports our body.

As we age, the collagen framework gradually deteriorates. The resultant deterioration causes the skin to become lax and wrinkled. It also loses its elasticity and resiliency. Collagen replacement therapy seeks to replace some of the lost or damaged collagen and thereby restores elasticity and structural strength to the skin. This replacement results in diminution of wrinkles and lines.

Collagen can be replaced topically by the use of collagen containing creams. This superficial treatment apparently only affects the epidermis, the outer layer of skin and its effects are temporary. Injections of medical grade collagen into the deeper, dermis layers results in a more profound and longer lasting effect. However, the benefits gained from collagen injections with substances such as Zyderm are also temporary and require repeated injections over time.

Telomerase and Younger Looking Skin

Utilization of the many and varied techniques available to rejuvenate a more youthful appearance of the skin may result in much more than simply a cosmetic effect. By strengthening the dermis against the most important factors that naturally age it, and by restoring damaged components to the skin, we may actually be causing it to "become" younger and more resistant to disease.

Telomerase therapy may be used to restore the genes of youth within each skin cell, and within each cell underlying the skin. In other words, telomerase therapy may not only allow the skin to look younger, but perhaps, genetically, to actually "be" younger.

The first and most famous scientific experiments that allowed researchers to realize that some natural substance (later discovered to be telomerase) was responsible for our cells having a limited lifespan, transpired several decades ago. Professor Hayflick, of Stanford University, discovered that cells of the type that support the skin structures, known as connective tissue cells, were only able to reproduce themselves a certain number of times. After this number of "rebirths" was achieved, the connective tissue cells could never again reproduce themselves and would grow old and die.

As the connective tissue cells in our subcutaneous tissues reach this "Hayflick limit," they cease to renew themselves and then slowly die off. As a result, our skin begins to sag and wrinkle. Telomerase skin therapy seeks to restore the ability of the skin cells to continually renew themselves, regaining and retaining their youthful vigor.

How to Rejuvenate the Skin

We all know that sunlight is a constant source of damage to the skin. A first step therefore is to protect the skin from damaging sunlight. But what simple steps can we take to restore the damaged components of the skin, particularly of our face? I believe that the dermal peel followed by regular administration of a combination of moisturizers and topical vitamins is best. This can be accomplished either by seeking care from your dermatologist, cosmetic surgeon, or aesthetician. Some kits for self-use at home are also available, but any type of dermal peel must be perfromed with caution to avoid a chemical "burn" injury.

Case Histories:

Here are two interesting cases in which the ancillary beneficial effects of natural treatments to treat systemic diseases resulted in cosmetic changes for the patient.

A gentleman I was treating several years ago with intravenous administration of detoxifying substances to treat his heart disease noticed a gradual and dramatic change in his hair color. It slowly transformed itself from gray, back to his natural dark brown hair color of youth. His hair also noticeably thickened. How did this happen? My guess is that the blood supply to the hair follicles was restored and this allowed the hair to become re-invigorated.

Another case involved a woman in her sixties who I was treating for colon cancer. She was treated a year ago, has no sign of colon cancer remaining in her system, and noticed another interesting effect. She and her friends and relatives saw that she began to look younger. She says her face, skin and nails seem to appear noticeably more youthful after the natural, holistic, cancer treatments. Oddly enough, when writing me to thank us for helping her fight the cancer, she gave us more praise for the unintended cosmetic response that occurred than she did for helping her with the cancer itself! Women really do prize their appearance!

Chapter Nine
EXERCISE AND DIET

Exercise

Proper exercise habits can be very useful to promote health and to delay the aging process. What type of exercise regimen is best suited for these purposes? There are many who believe that heavy exertional exercise is the most healthful and beneficial kind of physical activity, but this may not be true. Overly traumatic exercise involving heavy exertion may produce as many negative effects on the body as beneficial ones. In particular, overly heavy exercise can result in the body being presented with an abundance of free radical substances, which it must try to eliminate.

Free radicals are generated as a natural result of the process called “oxidative metabolism” in which oxygen “burns” food, "fuel" to produce energy. This exercise-induced abundance of free radicals can dramatically increase the aging process and is one of the prime reasons to encourage light to moderate exercises rather than heavy physical exertion exercises.

Would you say that your exercise regimens are primarily, 1) aerobic exercise that causes you to breathe heavily for at least ten minutes at each session, 2) non-aerobic muscle-building exercise such as weight training in which the large muscles are worked hard but the heart and lungs do not undergo much exertion, or 3) light physical exertion that is actually neither of the forgoing but does act to get the blood flowing and stimulates the muscles, heart and lungs to some extent?

Aerobic exercise consists of activities that cause your body to move at such speed that you are breathing heavily, your heart rate substantially rises, and you are consuming much larger amounts of oxygen than normal. This kind of exercise causes air and oxygen to be exchanged through the lungs and through the cells of the body. This is why it is called aerobic exercise. Running, dancing, very brisk and rapid walking, tennis, and rapid swimming are all examples of aerobic exercises. Heavy, non-aerobic exercises are those that primarily involve the use of the large muscles of the body but do not primarily cause the heart rate or breathing rate to increase significantly are non-aerobic physical exercises. Weight lifting is probably the most common form of this type of heavy non-aerobic exercise. Light, non-aerobic exercises include walking slowly or walking slowly with the dog, or doing light gardening.

It is recommended that you modify your exercise habits so that they include a predominance of aerobic activities along with some non-aerobic large muscle exercises. This combination seems to have the best anti-aging effects. Exercise can be a natural way to bolster the effectiveness of other anti-aging modalities. Exercise stimulates the body's tissues and cells to maintain them in a healthy manner.

As we exercise, many stimulatory signals are sent throughout the body. The stimulatory signals tell the tissues of the body that they are needed, and that the rest of the body will maintain them since they are needed. All of the cells of the body that are involved in the exercise process are therefore stimulated to maintain a more youthful condition by the mere fact that they are being used. In particular, this can have beneficial effects on the circulatory system.

HDL will maintain the blood vessels in a healthier condition, in part due to the cleansing of the lining of the blood vessels. HDLs or "high density lipoproteins" increase as a result of exercise and act to clean away fat deposits from the arterial walls. The "good" blood cholesterol "HDL" has been found by scientists to increase in response to regular exercise. This is one of the body's methods of cleaning out the existing arteries of atherosclerosis and other damaging blockages so that the cells and tissues that the artery supplies can be maintained in a healthy and active state. The total cholesterol is also known to decrease, as do harmful types of cholesterol such as LDL.

The circulation is also benefited by the development of collateral circulation, which involves new blood vessel development into areas formerly served only by older vessels. This provides an area of body tissue with two sources of oxygen and nutrient-rich blood. Both the new and the old vessels are available to perfuse the tissue. This is especially beneficial if the original blood vessel is ever damaged or blocked. In a heart attack, a blood clot or "thrombus" blocks the original blood vessel. The collateral blood vessel can supply the same area and maintain the health and viability of the tissue being supplied with blood, protecting the area from damage. Another important benefit of the proper amount of exercise is that human growth hormone can be induced and its production by the body increased.

The immune system is also benefited by exercise as long as the exercises do not overly exert the body. Overly heavy exercise can actually stunt the immune system. Athletes who are in very heavy training regimens have been found to exhibit signs of immune system dysfunction and decreased immune system activity due to this overly burdensome exercise.

An Anti-Aging Exercise Program

A life extension exercise program, consisting primarily of aerobic exercise along with some large muscle exercise accomplishes several goals. The aerobic exercise improves circulation, strengthens the heart and improves elasticity of muscles, tendons, ligaments and joints. The muscle training exercise increases lean muscle mass, diminishes body fat, and stimulates the natural production of human growth hormone.

Both the aerobic and the muscle training exercises act to stimulate the tissues of the body to remain alive and functional. Lack of such regular exercise tells the body that the heart, lungs and muscles are not needed. This is one of the ways in which lack of exercise leads to atrophy, which is a gradual wasting away of the heart, lungs and muscles. Sadly, as these beneficial tissues are gradually lost due to lack of exercise, they are commonly replaced with fat.

It may be wise to have an exercise program of at least three to five days a week. A regular schedule such as Monday, Wednesday and Friday, or Monday through Friday will be most beneficial. It might be wise to leave the weekends free of planned exercise for other activity and travel. Thirty to sixty minutes a day, three days a week, is adequate exercise for the purposes of a life extension program. For example, in a forty-five minute period, roughly thirty minutes should be devoted to aerobic exercise and fifteen minutes to muscle training.

What is the purpose of the aerobic and muscle training portions of the exercise plan? The aerobics portion gives several benefits to longevity. First of all, since aerobic involves utilizing the heart, lungs and blood vessels to a significant extent, all of these tissues will be benefited. The circulation will improve your ability to absorb oxygen and the release of gaseous waste products will be stimulated. Muscle and other body tissues will benefit from the rapid movements involved in aerobic exercise.

So, what should the thirty minutes of aerobic exercise consist of? It should consist of any type of aerobic activity that you find pleasurable. For example, a thirty-minute brisk walk, a thirty-minute combination run or jog with brisk walking, or thirty minutes of using an exer-cycle or treadmill.

Other types of exercise equipment that also suffice are those in which your arms and legs are kept moving. With these you are induced to breathe more rapidly, causing your heart to pump much faster than normal. This aerobic exercise should not be overdone and the exercise program should be started slowly and carefully. For example, if your goal is to eventually exercise thirty minutes aerobically three days a week, it would be wise to start out with five to ten minutes of aerobic exercise and increase it to thirty minutes over a period of months. This will depend on your physical attributes and level of training at the initiation of the exercise program.

How do we know what level of aerobic exercise one should attain? I believe that for purposes of life extension, one might choose to use the method of maximal heart rate. If we use the formula 220 minus the age, we can calculate a maximal exercise heart rate. So, for a seventy year old, the maximum exercise induced heart rate of 150 beats per minute would be tolerated after the body has become used to regular exercise.

We would like to exercise up to a range at which our heart is beating at 70 to 80 percent of the calculated maximum predicted heart rate. We will try not to exceed the maximum predicated heart rate. For example, the seventy year old man would slowly increase his exercise tolerance so he could do aerobic exercise for a period of thirty minutes, maintaining 120 beats per minute, with maximal heart rate not exceeding 150 beats per minute. This is calculated by the formula "220 minus his age (70)," and yields the top maximal exercise heart rate of 150. Eighty percent of the 150 is 120 beats per minute, which is a good range to choose for a majority of your thirty-minute period.

Can we simply jump right from rest into thirty minutes of aerobic exercise? No, it is not healthy. The aerobic portion of your exercise should be preceded by a warm-up and stretching period. The warm-up can start with stretching of the muscles and tissues of the arms, legs and upper body including the neck. You might then choose to engage in a light warm-up activity such as sit-ups or jumping jacks.

Any warm-up activities should be done in a progressively rapid manner, starting from a fairly slow pace.

Another form of warm-up would simply be to begin, after stretching, whatever aerobic exercise you choose to do. Begin it at a much slower pace, gradually building to the more rapid pace that will yield your exercise heart rate of 80 percent of the predicted maximal exercise heart rate.

Following the aerobic portion of your exercise, it is beneficial (when you can tolerate it) to have a ten to twenty minute period training the large muscles. What we would like to do in this segment of the exercise session is cause the large muscles of the body to be exercised in a fairly non-aerobic manner. In other words, in a manner in which their strength, tone, and perhaps size will be increased due to the exercise.

Muscles can be used in different ways. There are two primary ways in which we tend to use our muscles during exercise. During aerobic exercise we tend to move the body in rapid short bursts in which not a great deal of force needs to be exerted by the muscles. For example, in using a treadmill, our arms are moving rapidly as we walk. However, this does not require the arms, chest or upper back muscles to do anything but move a light weight, i.e., your arms, in a rapid and rhythmic fashion. Even the legs are not performing nearly the heavy work of which they are capable. For example, while we walk or run our leg muscles are required to rapidly project our own body weight forward. However, they are designed to lift far more heavy burdens than that.

The opposite of aerobic muscle exercise occurs when the muscles are called upon to work for shorter periods of time, but at a much greater level of exertion. Lifting weights is an example of such heavy exertion exercise. In lifting weights, or using resistance exercise equipment, the muscles must work for shorter periods of time but to a much greater exertion. In doing forearm curls and biceps curls, the muscles of the upper arms, depending on the weight, can be exercised to the maximum exertion that they have the strength to perform. This causes the muscles to enlarge and it causes the body to metabolize the fatty tissues in favor of developing muscular tissues.

The goal of the large muscle training portion of your exercise time then would be to specifically exercise the large muscles of the upper and lower body, including the legs, arms, chest and back to produce greater strength and muscle mass in these areas. This is also true for women. Women will develop a leaner, healthier body, which has a greater tendency toward longevity if this part of the exercise program is included along with the aerobics.

How can training of the large muscles be accomplished? We can use simple and inexpensive methods to achieve exercise of the large muscles. This can be accomplished in several ways. One of the simplest ways would be dumb bells and strapped-on weights. For example, one or two sets of dumb bells can be used to effectively exercise the upper body and arms. Weights, strapped onto the ankles, can be used to effectively add weight stress during exercise to the leg muscles thereby giving the leg muscles a more pronounced workout. Other methods would include certain types of exercise machines that specifically target the use of large muscles as you use the equipment. Nautilus, Cybex, and other weight-training machines are examples of this type of exercise equipment.

Diet

A few simple but essential concepts underlie a basic rational anti-aging dietary strategy. The essential concepts are: 1) avoid fat, 2) enjoy unlimited proteins, 3) avoid foods that are deep-fried, seared at high heat, or cooked in significant amounts of fat, and 4) include ample whole uncooked fresh fruits, vegetables and whole grains at each meal.

Let's first discuss essential concept number one. It is the avoidance of fat in the diet. Dietary fat can come in many forms. In the common American diet, one of these forms may be as "hidden fats." Fats can hide in egg yolks, cream, cheeses, sauces, condiments, mayonnaise and most notably salad dressings.

Fortunately, American supermarkets have recently begun to offer non-fat versions of almost all of these things. Fats that we can all recognize would be those present in fatty meats. For example, hamburger, sausage and any meat that contains visible fat is obviously high in fat content and should be avoided. However, leaner red meat that has minimal marbling can be enjoyed. For example, certain types of flank steaks, rib eye steaks, and London broil, when they have the extraneous fat trimmed off, are low enough in fat that they can be enjoyed.

You may ask yourself, should I try to count my fat calories? Or count my total calories? Or, in some other way try to quantitatively analyze how much fat I have in my diet? The answer to that is no. It is more reasonable to simply develop habits of eating foods that in general are healthy rather than worrying about what percentage of your total daily calories come from fat, protein or carbohydrates. However, as a general rule of thumb, if a person were to analyze their calorie intake, a prudent longevity diet would include less than ten to fifteen percent of the total daily calories from fats.

Are all fats bad? Not at all. Certain fats are essential. We must have them or our body will become ill and deteriorate. Most modern people have such a high fat diet that even if they take arduous measures to limit their fat intake they are probably not going to limit it to such a severe extent that they fail to have adequate fat in the diet.

There are certain foods, fairly high in fat, which we are going to encourage you to enjoy on a limited basis. These foods contain fats that are either essential fats, or fats and oils that are very beneficial to the body. In particular, we are talking about deep-sea cold water fish and certain plant-derived oils. Salmon is a good example. Salmon is a fairly high fat animal flesh. However, it contains omega-3 fatty acids, which are very beneficial to our body and may have significant arterial anti-clogging effects. Likewise, certain oils should be used in cooking and in the preparation of salad dressings and condiments. These include fresh borage seed oil, flaxseed oil, olive oil and safflower oil.

Flaxseed oil is a source of beneficial omega-3 essential fatty acids. Borage seed oil has other (omega-6) essential fatty acids, which complement those found in flax oil. Mixtures of flax and borage seed oils can be a valuable addition to our anti-aging diet. Together, flax and borage seed oils yield a significant amount of both the omega three and omega six essential fatty acids. Olive oil also has been found, due to its predominately monosaturated fat content, to be much healthier than more commonly used oils such as corn oil, butter and lard.

When it comes to fats, it is wise to avoid processed fats. Transfatty acids are man-made products of oils and fats that have been produced to add palatable texture and consistency to processed foods. They are present in a very wide variety of processed and pre-packaged foods.

Many times you may notice that they have been included in the packaged foods by reading the label. It may say, "partially hydrogenated vegetable or other oils." If it says, "partially hydrogenated," this is an indication that transfatty acids are present in the food. The transfatty acids may, in the future, be shown to be deleterious to health, and in particular may be detrimental to longevity. It is postulated that their effects in disrupting cell membranes may limit a healthy lifespan in humans.

Let's talk now about point number two, which is the enjoyment of unlimited amounts of carbohydrates and proteins. There are three major constituents of foods. They are fats, proteins and carbohydrates. I think we all know what fats and oils are. Proteins are substances that make up the muscular flesh of animals and are found to some extent in almost all plant food but are particularly rich in beans and other legumes. Carbohydrates are sugars, starches and related substances found predominately in plant-derived foods.

Once again, rather than worrying about how many grams of fat, protein or carbohydrates we ingest each day or about the total percentage of our daily calories, it is better to develop simple, easy-to-follow habits in which we systematically avoid fatty foods and enjoy unlimited quantities of healthy, plant-derived carbohydrates and protein foods to satisfy our hunger.

Under what conditions should carbohydrates or proteins be limited or their intake considered more carefully? High protein intake should be avoided by people with kidney disease. So, if your physical exam, past medical history, or lab tests indicates kidney disease, you should consult your physician about your intake of protein. Kidney disease might be indicated on your chemistry panel by elevated BUN and creatinine.

Likewise, those people who have elevated blood sugar or diabetes should more carefully consider carbohydrate intake. Once again, your physician and/or dietician should be an integral part of your dietary planning. This is particularly true when it comes to carbohydrate intake if you are diabetic or have a significant pre-diabetic disposition, or if you suffer from kidney disease.

Enjoying a diet that is filled with ample whole, uncooked, fresh fruits, vegetables and whole grains is an essential concept. We are all aware that the modern diet contains far less fiber than diets of humans in past generations. Processing and commercial preparation of foodstuffs removes much of the natural fiber and has led to increased blood cholesterol and cardiovascular disease.

It is even implicated in the propensity of certain individuals to develop some cancers. From the perspective of anti-aging, another benefit from fiber intake is noted, which is the effect of dietary fiber to decrease the transit time of the food as it passes through the digestive tract. The resulting slower digestion and absorption of nutrients allows for less severe blood sugar spikes.

If blood sugar (glucose) rises rapidly, insulin is released to deal with it. This combination of glucose and insulin can have detrimental effects on the arteries and other tissues of the body. Insulin can have direct caustic effects on tissues, and is thought by some aging researchers to be one of the prime causes of the aging process. Likewise, glucose can have its own damaging effects on tissues. "Advanced glycosylation end products" or "AGEs" are one such class of glucose-related byproducts that may damage and age our bodies. AGEs occur when glucose interacts with proteins and fats. Much study is presently being devoted to AGEs by today's biogerontologists.

Undamaged nutrients, in their natural form, can best be derived from whole, uncooked and unprocessed plant foods. Such plant-derived nutrients or "phytonutrients" are novel sources of the minerals, vitamins, and other, as yet undiscovered, beneficial elements crucial to optimum health and longevity.

When plant foods are consumed whole and unprocessed, they impart to us all of their nutrients bound together in ways that may make them more powerful sources of nutrition than each nutrient would be individually. These benefits are difficult to obtain in foods that are cooked or processed.

Achieving and maintaining an optimal dietary routine is extremely beneficial to the anti-aging program and to good health in general. There are several benefits to eating right. One of these is that your body weight will slowly change towards what is called an "ideal body weight."

Your calorie intake and ideal body weight is one of the best understood and studied factors effecting longevity. Whether too heavy or too light, any significant deviation from the norm could shorten your lifespan. Studies have shown conclusively that being significantly underweight as well as being overweight is correlated with shorter lifespan and more physical ailments during the lifetime. As your dietary and exercise regimen is modified, you will slowly begin to notice that adipose tissue (fat) will tend to leave your body while lean body mass, muscle, bones, and connective tissues will remain or become more prominent. This leaning of the body will be caused by several different factors, which include proper dietary habits, an adequate exercise regime, hormonal re-balancing, and richer nutrients through supplementation.

Dietary Trans Fats

Is it less healthy to consume man-made dietary fats? Do natural plant-derived fats such as those found in grains and nuts aid our longevity in comparison to chemically transformed fats? These questions are considered in this section, which discusses research on a type of chemically altered man-made fat known as trans fats. Trans fats are very common in our diet as they are used extensively in the manufacture of fast foods and packaged foods.

Multiple factors are causing humans to ingest greater quantities of trans fatty acids. Potential deleterious effects of these dietary transfats include altered aging, serum lipoprotein changes such as high cholesterol, increased blood clotting or "platelet aggregation," spasms of the arteries supplying blood to the heart muscle or "coronary vasoconstriction," and malignancy.

Throughout the past several decades, American consumers have been encouraged to purchase more products containing dietary polyunsaturates in place of saturated fats. In order to offer a greater variety of polyunsaturated foods, producers chemically transform a portion of polyunsaturates into the trans fatty acid form. Consequently, the dietary intake of trans fatty acids has increased.

The known potential deleterious effects of these dietary trans fats are many and include serum lipoprotein changes, platelet aggregation, coronary vasoconstriction, and malignancy. The scientific investigation of the role of trans fats in human disease and risk factors for disease is at present inconclusive. However, they are intriguing and point out our need to gain a better understanding of the potential role of dietary trans fatty acids in human disease.

Encouraged to consume less saturated fat and replace it with polyunsaturates, the American population has dramatically increased its use of vegetable oils. In addition to naturally occurring vegetable oils, our diets also increasingly consist of the man-made products of industrialized food production, the hydrogenated oils (such as trans fats).

The partial hydrogenation of polyunsaturated fats from vegetable sources yields many oil products useful in food processing. It is estimated that fifty percent of the vegetable fats consumed in the American diet have been processed via partial hydrogenation into margarine, shortening and oil components of prepared foods.

Hydrogenation alters the texture of foods in salutary ways. Margarine is a spreadable form of vegetable oil which, being solid at room temperature, is more convenient than butter, which partially melts at room temperature. Partially hydrogenated vegetable oils impart other properties to foods such as increasing the flakiness of piecrusts and the creaminess of puddings. These hydrogenates are now at the core of a debate concerning potentially deleterious effects caused by these products of modern food processing.

Industrial partial hydrogenation converts naturally occurring polyunsaturated oils into various chemical forms. Among these chemical forms of oils are fatty acid isomers. Isomers are subtle differing forms of the same chemical compound. The compounds each contain the same type and number of atoms. Although their atoms are arranged in essentially identical ways, they differ in subtle variations of the directional position of some atoms in relation to their counterparts. These subtle differences in positioning can result in significant differences in chemical bonding forces within the compound.

The normal, natural isomeric form of partially or polyunsaturated fatty acids is the "cis" form, in which hydrogen atoms are located on the same side of the adjacent carbon-carbon double chemical bond, while in the "trans" version, the hydrogen is found on opposite sides of the double bond. This seemingly insignificant change in spatial positioning causes the fatty acid chain in which it is contained to take a different folding shape. The normal "cis" isomer folds back upon itself while the "trans" isomer occurs as a straight, linear molecule.

Shape differences between the two forms yield different physical properties in prepared foods. The trans fatty acids are more stable and durable, resisting deterioration of the product due to temperature variations and other factors. The question being researched by science, therefore, is whether these differences in physical characteristics, which cause beneficial effects for food producers, may cause adverse physical effects to human consumers when incorporated into the body after ingestion.

Diets rich in trans fatty acids of partially hydrogenated vegetable oils have been shown to raise low density lipoproteins, while also lowering high density lipoprotein (Mensink & Katan, 1990). Other studies have shown elevations in total serum cholesterol and in the triglyceride level of the blood in humans (Erickson, 1964) (Ahrens, 1957) (Anderson, 1961). Researchers have also found similar results when studying animals (rabbits and swine) fed diets rich in trans fatty acids. Specifically, the animals were found to develop increased total cholesterol and LDL, along with lowered levels of HDL and coronary artery prostacyclin (Elson, 1981) (Royce, 1984) (Kritchevsky, 1982).

Coronary artery prostacyclin (Prostaglandin I2) is produced by the endothelial cells, which line the coronary artery walls, and has two beneficial effects. It acts to inhibit platelet aggregation, particularly after acute damage to the endothelium, thereby decreasing the opportunity for a blood clot to block the heart's blood supply by coronary thrombosis. It also acts as a vasodilator, causing increased coronary artery diameter resulting in improved blood and oxygen supply to the heart muscle. Loss or diminution of coronary artery prostacyclin may therefore pose a significant risk factor, in and of itself, for acute and chronic cardiac disease.

Studies on cholesterol and heart disease have yielded information teaching us which fats to avoid in our diet. Recent advances in the elucidation of lipid metabolism and atherogenesis point to the total daily intake of saturated fatty acids as the prime culprit in coronary and peripheral artery disease. This is in contrast to the former, long-held opinion that dietary cholesterol serves as the most significant factor. It appears that the availability to transport cholesterol as LDL is limited by the bioavailability of enough saturated fatty acids to produce serum lipoproteins such as VLDL and LDL.

Cholesterol metabolism also seems to be regulated more profoundly by liver biosynthesis than by sterol absorption via the digestive tract. An important corollary to the role of saturates is the phenomenon of their partial replacement in the diet by trans fatty acids. The ad campaigns of the past many years touting the benefits of margarine over butter are a prime example of the basis for this displacement phenomenon. Due to the similarity in physical characteristics and cooking propensities, margarine and other hydrogenates tend to replace saturates instead of polyunsaturates or monounsaturates.

A study attempting to clarify the actual dietary effects of trans fatty acids in the American diet on blood lipoprotein levels revealed several pertinent findings (Judd, 1994). The dietary trial of cis and trans mono-unsaturated fatty acids and saturated fatty acids employed 29 male and 29 female human subjects and was six weeks in duration. Results showed that compared to the cis isomers, transfats produced a significantly greater serum LDL, with saturates producing an approximately one percent greater rise in LDL than did the trans fats.

HDL levels were also shown to decrease when the subjects consumed the high trans fat diet. Conversely, the high saturated fat diet caused a higher level of beneficial HDL, which may in some degree mitigate the deleterious effects of its propensity to raise LDL levels. Diets high in trans fatty acids, therefore, were shown to produce two negative factors simultaneously. Beneficial HDL was lowered while harmful LDL was raised.

Saturated fats, which are being displaced in the American diet by trans fats, contrast with the actions of trans fats in that they simultaneously raise harmful LDL and beneficial HDL. It seems likely that the actual effects of this displacement phenomenon are much less than those found in this study because it is estimated that the average American diet has not yet reached levels of trans fats used to produce these findings. The study showed that lower, more moderate levels of trans fats, of greater similarity to the make-up of the current American diet, yielded more salutary results. These more moderate levels of diet trans fats revealed in a smaller elevation in LDL and no significant change in HDL as compared to the diet richer in cis isomers.

This dietary trial seems to serve as an example of the better methodologies available to discern risk factor effects. Larger groups of subjects, from varying ethnic backgrounds, studied over a longer time period, following more research into the trans components of the actual American diet, may yield more definitive results.

Another interesting question concerning dietary trans fatty acids is whether they exhibit a propensity to stimulate endothelial and vascular smooth muscle cell proliferation, thus compounding and exacerbating the growth of atherosclerotic plaque. Other mitogenic and growth factors have been implicated in the development of atherosclerosis. Included among these are platelet-derived growth factor and fibroblast growth factor.

The cell membrane is composed primarily of lipids. Proteins, cholesterol, and other constituents are also present. It is the function of the cell membrane to regulate the intracellular environment, and to allow signal transduction from the exterior to cause an intracellular effect. Many of the oncogenes encode cellular membrane signal transduction proteins and are thought to play important roles in oncogenesis when they participate in deranged signaling.

It remains to be seen if unusually configured fatty acid moieties also play a role in derangement of the cell membrane's ability to properly transduce signals and maintain homeostasis in the intracellular compartment. If unusually configured fatty acids such as trans fats are incorporated into cells, it would be important to discover whether or not they interfere with normal cellular functioning.

Evidence correlates the incorporation of significant fractions of trans fatty acids into human cells with the relative percentage of these isomers ingested from the diet (Hudgins, 1991). The Hudgins study showed that the trans isomers incorporate into adipose tissue but failed to show a strong tie between trans fatty acid incorporation into adipose tissue and clinical risk factors for cardiovascular disease.

Another study at the University of Southern California (London, et. al) failed to show any significant relationship between gluteal adipose tissue, trans fatty acids and breast cancer or benign proliferative breast disease. Further work to elucidate fatty acid isomer incorporation into other types of cells may be beneficial in determining whether they play a direct cell proliferative role in atherogenesis or cancer.

In conclusion, many potential harmful effects from the incorporation of trans fats into the modern American diet have been proposed. Studies in humans and animals point to the possibility that several of these effects, including cardiovascular disease risks, may indeed occur. However, at present, the result appears inconclusive for several reasons. The first reason is that there is a question as to whether the finding from animal models are directly extrapolatable to human populations. This is a question often posed when non-human subjects are employed in an attempt to elucidate human disease processes. Secondly, we need to gain a better understanding of the actual amounts and types of trans fats presently in the typical American diet. Finally, prospective trials, utilizing larger groups and conducted over an extended time period, may prove more useful in understanding the role of dietary trans fats.

How to Eat for Health and Longevity

Perhaps the easiest sign of a poor diet and exercise "lifestyle" is obesity. So many today are fat and flabby. Not only do we wish to be leaner so that we will look more stylish, but becoming leaner often confers many health benefits as well. A simple diet and exercise plan that virtually anyone can use consists of just a couple changes in our normal routine. We must make time to exercise every other day for at least twenty minutes. Walking, running, lifting weights, aerobicycles …anything that is fun and gets the heart pumping will suffice.

Changing the diet is just as easy. We must first begin by refusing to purchase anything but lean and wholesome foods. Exclude all junk foods, fast foods and fried foods. Buy whole grain cereals for breakfast. Choose lean meats, fish, and poultry as main courses. Add plenty of fresh vegetables with the main course and ripe fruit for desert. This is simple, healthy, and easily attainable by anyone reading this book.

In my experience, these simple and non-intrusive changes often lower cholesterol, cause weight to drop and energy to rise. Some may say, "What about a healthier, stricter diet and way more exercise, isn't that even better?" Perhaps, but this is a simple regimen that anyone can find acceptable and do-able. Doing something, getting started, taking that first step towards a healthier lifestyle is the most important thing!

Case History:

Recently, I counseled a 52 year old woman regarding her diet and her concerns over recent weight gains. She decided to try my written weight loss and physique enhancement program which I have named my "BioThin Program." It consists of a guidebook and supplements to assist in the initial weight loss process, and then to extend the physique enhancement into a life-long habit.

She lost six pounds in less than two weeks, and has since continued to slowly gravitate towards a permanent "set-point" weight which her body will choose for her. This process of healthy eating and sensible exercise allows our own body to decide on the correct weight for each of us.

References

Mensink, R. P. and Katan, M. B., “Effect Of Dietary Transfatty Acids On High-Density And Low-Density Lipoproteins Cholesterol Level In Healthy Adults,” *New England Journal of Medicine* 323 (1990): 439-445.

Erickson, B. A., Coots, R. H., Mattson, F. H., Kligman, A. M., “The Effect Of Partial Hydrogenation Of Dietary Fats, Of The Ratio Of Polyunsaturated To Saturated Fatty Acids, And Of Dietary Cholesterol Upon Plasma Lipids In Man,” *Journal of Clinical Investigation* 43 (1964): 2017-2025.

Ahrens, E. H., Hirsch, J., Insull, W., Tsaltas, T. T., Blomstrand, R., Peterson, M., "The Influence Of Dietary Fats On Serum Lipid Levels In Man," *Lancet* 1 (1957): 943-53.

Anderson, J. T., Grande, F., Keys, A., "Hydrogenated Fats In The Diet And Lipids In The Serum Of Man," *Journal of Nutrition* (1961): 388-94.

Elson, C. E., Benevenga, N. J., Canty, D. J., et. al., "The Influence Of Dietary Cis And Trans And Saturated Fatty Acids On Tissue Lipids Of Swine," *Atherosclerosis* 40 (1981): 115-37.

Royce, S., Holmes, R. P., Takagi, T., Kummerow, F. A., "The Influence Of Dietary Isomeric And Saturated Fatty Acids On Atherosclerosis And Eicosanoid Synthesis In Swine," *American Journal of Clinical Nutrition* 39 (1984): 215-22.

Kritchevsky, D., "Transfatty Acid Effects In Experimental Atherosclerosis," *Fed Proc* 41 (1982): 2813-17.

Judd, J. T., Clevidence, B. A., Muesing, R. A., Wittes, J., Sunkin, M. E., Podczazy, J. J., "Dietary Transfatty Acids: Effects On Plasma Lipids And Lipoproteins Of Healthy Men And Women," *American Journal of Clinical Nutrition* 59 (1994): 861-68.

Hudgins, L. C., Hirssch, J., Emken, E. A., "Correlation Of Isomeric Fatty Acids In Human Adipose Tissue With Clinical Risk Factors For Cardiovascular Disease," *American Journal of Clinical Nutrition* 53 (1991): 474-82.

London, S. J., Sacks, F. M., Stampfer, M. J., Henderson, I. C., Maclure, M., Tomita, A., Wood, W. C., Remine, S., Robert, N. J., Dmochowski, J. R., et al., "Fatty Acid Composition Of The Subcutaneous Adipose Tissue And Risk Of Proliferative Benign Breast Disease And Cancer," *Journal of the National Cancer Institute* 85(10) (May 19, 1993): 785-93.

Chapter Ten
Melatonin And Sleep Therapy

Humankind has sought immortality and eternal youth since the dawn of history. Central to achieving this desire has been the necessity to understand, or explain through cultural myths, the underlying reasons for, and mechanisms of, the aging process. Even in today's seemingly advanced world of technology and biomedical science, the underlying mechanisms of aging remain a mystery. This chapter will explore one potential factor in the natural aging process. That factor is the central nervous system, melatonin, and its precursor, serotonin.

Many theories to explain the underlying cause of normal aging have been proposed. These include theories implicating central nervous system degeneration as the fundamental root cause of aging. The central nervous system is recognized in biomedical science as the main control point and regulator of the body's complex physiological processes through its network of peripheral nerves and via the hormonal system. This fact, along with the apparent highly predictable timing and predictably sequential nature of human senescence, leads many to conclude that aging is not a random accumulation of physiological insults, but is rather a planned and possibly pre-programmed process.

Further points of circumstantial evidence pointing to pre-programmed senescence caused via centralized control of the underlying normal aging process are the apparent human maximum life span, now thought to be one hundred fifteen years, and the rectangularization of mortality curves in advanced western societies.

Melatonin, and in particular pineal melatonin, may serve as the underlying mediator of the central nervous system's control over the body's processes of normal aging. It has long been postulated that the pineal gland, located deep within the brain, is the body's directing gland controlling timing, onset, and duration of the many developmental stages producing a mature, reproductively capable human being. It is now becoming evident that besides controlling our growth and development from childhood through adolescence and into adulthood, this gland may also serve as the biologic "time clock" regulating the sequence of events in late adult life that lead to senescence and eventually death.

This "time clock" phenomenon may occur via the controlled production and degradation of melatonin and its precursor serotonin by the pineal (Klein, 1978, Klein and Weller, 1970). What sets and controls the pineal time clock? Perhaps the telomere time clock of the pineal cells? Let us explore the phenomenon of melatonin further.

The pineal appears to utilize a ubiquitous environmental cue, daytime sunlight and the nocturnal absence of sunlight, as the basis of the gland's time clock mechanism. This incessant pattern of day and night sunlight cycles seems to serve like a metronome setting the rhythm for the pineal gland's timing system. It may be that even as the battery driven second hand on a wristwatch steadily ticks off the minutes and then hours on the clock-face, so too the pineal gland serves as our life clock as it ticks off each day-night cycle.

Another possible mechanism enabling the pineal to "keep time" and regulate our growth and aging is that the gland senses the degree of its own gradual calcification. The pineal normally slowly becomes calcified as each day passes (Welsh, 1985). Assuming this rate of calcium deposition to be steady, the pineal may be able to utilize measurements of its own calcification to determine where the organism is within the life-cycle, and release the proper amount of melatonin and other substances appropriate for that point in the life cycle.

But what exactly regulates the calification of the gland? It seems plausible that the true timing mechanism may be the planned senescence of pineal cells based on the genetic time-clock we call the "telomere clock." As the telomeres of various pineal cells slowly degrade, the pineal cells may die in an orderly fashion, and subsequently become calcified. This build-up of calcified dead cells may serve as the timer for the changes in melatonin release by the gland.

As the melatonin release pattern changes with age, it then signals the rate of aging for the entire organism. Release of pineal gland products such as melatonin into the bloodstream may therefore signal the individual cells of our body as to their age and the level of function expected of them. An old pineal gland, therefore, may send less total, and especially less nocturnal, melatonin into the blood stream. This may then cause the body cells to change to a senescent condition or die.

The pineal, along with another area of the central nervous system which activates the pineal, the hypothalamic suprachiasmatic nucleus, measures and is synchronized by the day-night cycle, and also senses and responds to physiological states in the body through the gland's rich blood vessel supply. The resulting hormonal control of physiologic processes by the pineal gland is mediated via the gland's secretion of several substances.

These substances include melatonin, serotonin, other related substances known as methoxyindoles, and peptides such as arginine vasotocin. The brain's integrative control over widespread physiological activities of the body is mediated by the pineal gland's precise secretions of exacting quantities of these substances (Klein, 1978) (Klein and Weller, 1970).

Of the various substances secreted by the pineal gland, melatonin seems to have certain unique anti-aging and restorative effects. Higher levels of melatonin are produced at night. At the same time, lower levels of serotonin, thought to have pro-aging qualities, are produced nocturnally (Rozencweig et al, 1980). During daylight hours a greater amount of serotonin is produced as compared with melatonin.

This interaction of melatonin and serotonin ensues from the pineal's production at night of melatonin from its serotonin precursor. Melatonin, the predominant nighttime hormone, may serve to counteract the sequelae of serotonin, the predominant daytime hormone. A greater amount of serotonin during the active, daylight hours may serve to enhance increased metabolism, growth, maturation, and reproductive activities. A greater prevalence of melatonin during the restful nighttime hours may produce repair and rejuvenation of the damaging results of daytime serotonin. These potentially damaging by-products of daytime serotonin-driven anabolic activities may include various waste products and cellular damages.

The resultant effects of melatonin's rejuvenating properties may add up to longer life span and lessened rates of malignancies. It is postulated that an essential component of each night's rest is a partial melatonin mediated restoration of the many pro-aging effects of daytime serotonin, such as metabolic waste by-products and residual effects from cellular oxidation. A gradual breakdown of this restorative system following early adulthood may lead to the eventual signs of senescent deterioration. Pineal gland production and secretion into the general bloodstream of melatonin and its precursor serotonin may serve to set our life-clock and determine longevity.

Why does the central nervous system allow melatonin levels to fall and aging to occur? Perhaps the pineal has a set program to end the individual's life, just as it had a set program for the individual's initial growth and maturation. Or maybe the pineal has a set program that takes the organism up to the point of sexual maturity but does not have a program to use thereafter, resulting in a gradual decline in its role as a master regulator. Likewise, perhaps the pineal itself ages or some combination of the foregoing causes the pineal to allow the aging process to ensue.

Pineal gland cells and other brain cells are known to age and die. Like most neuronal cells, pineal cells cannot regenerate or replicate themselves. Loss of pineal cells due to amyloid and lipofuscin accumulation (waste products of cellular metabolism), gliosis (entanglement), loss of nervous stimulation by the sympathetic nervous system, nondegenerative calcification, and loss of elasticity leads to a gradual deterioration in pineal function resulting finally in total pineal failure.

Melatonin and the biochemical enzyme which produces it from serotonin (serotonin N-acetyl transferase) have been shown to decline as the gland ages, as does the ability of the gland to produce nocturnal elevations in melatonin (Mirmiran et al., 1984) (Iguchi et al., 1982). Consequently, pineal aging results in less pineal and circulating melatonin and an increased level of serotonin relative to the level of melatonin. Relative loss of anti-aging melatonin may lead to the gradual physical and mental deterioration that is part of the human aging process.

Further proof of the role of the pineal gland in aging comes from laboratory experiments in which animals who have had their pineal glands damaged or removed show premature signs of aging. In these experiments, it was found that giving evening melatonin supplementation (exogenous melatonin) counteracted the aging processes and resulted in a prolonged life span for elderly mice (Pierpaoli and Maestroni, 1987).

In similar fashion, surgical implantation of pineal glands from young mice to elderly mice resulted in decreased signs of senescence and a prolongation in life span for elderly mice (Pierpaoli, 1991). Researchers have also found that drugs which enhance the action of melatonin, and drugs that stimulate melatonin production also cause increased life spans in these animals (Anisomov et al., 1989).

Endocrine And Immune System

Two other master control areas of the brain, the hypothalamus and pituitary, are regulated by pineal melatonin production. In this manner, the body's entire endocrine system and the hormones it releases are influenced by the pineal gland. The hormonal regulatory system may prove to be a critical component of the natural aging process in that many of the hormonal changes of late adulthood are directly correlated with the signs and dysfunctions of aging.

Researchers have uncovered indications that melatonin directly influences thyrotropic-releasing hormone (TRH) and thyroid stimulating hormone (TSH) (Pierpaoli and Yi, 1990). TRH and TSH are important signaling hormones secreted by the hypothalamus and pituitary. TRH and TSH influence the production of thyroxin. Thyroxin, the primary hormone responsible for setting our metabolic rate, has been shown to increase if melatonin is administered to the test subject on the prior night.

Other endocrine organs influenced by melatonin include the adrenal glands, which in rats become enlarged after pinealectomy (Wurtman et al., 1967) (Jouan and Samlirez, 1964). Increased aldsosterone raises blood pressure, a common problem among the elderly, which may in turn damage blood vessels and brain cells. Other common ailments of the elderly, including strokes and heart attacks, are related to elevated blood pressure.

One of the most dramatic occurrences in female aging is menopause, which is caused by the discontinuance of female hormone levels present during the youthful reproductive years. Menopause signals the loss of reproductive capability, demineralization and loss of bone strength, skin changes, vaginal and urinary tract atrophy, thermoregulatory dysfunction (hot flashes) and mood or behavioral disturbances.

Production of female hormones is controlled via the brain's secretion of pituitary luteinizing hormone (LH), and follicle stimulating hormone (FSH). Melatonin depresses pituitary LH and FSH production, thereby decreasing female hormone production. Animal experiments tend to reinforce the theory that melatonin dramatically influences female reproductive system aging. Rats that were administered nocturnal melatonin via their drinking water exhibited marked delayed aging in their reproductive functions and hormone levels (Trentini, et al., 1992).

Human growth hormone and insulin production are also influenced by melatonin (Smyth and Lazarus, 1974). Circulating levels of human growth hormone are thought to markedly influence the relative prevalence in an individual of many of the changes seen in aging such as bone and muscle loss, fat gain, and skin fragility. Insulin, chemically similar to growth hormone, is also regulated by melatonin. Changes in insulin and blood sugar levels are common in the elderly.

The nervous system and pineal gland may well influence, through varied control mechanisms, both the immune system and the endocrine system. Melatonin particularly seems to play a role in modulating immune system capabilities. Experiments with rodents have shown that melatonin can protect against potentially fatal viral infections (encephalomyocarditis virus), and increase antibody production (Pierpaoli and Maestroni 1987).

Melatonin has also counteracted experimentally induced immunodepression caused by high doses of steroids via improving antibody production and thymus gland mass (Maestroni et al., 1986). Interleukin-2, often in the news due to its anti-cancer and anti-infection capabilities, is also influenced by melatonin (Lissoni et al., 1990).

Sleep

One of the most common complaints of elderly individuals is sleeplessness. They complain that their sleep patterns are disturbed and that they do not feel as rested and refreshed after sleep as they did at a younger age. Melatonin has been shown in scientific studies to influence sleep and other functions of the brain including epilepsy.

Melatonin synchronizes EEG patterns, stabilizes the brain's electrical activities, and seems to have a beneficial effect on jet lag (Brailowsky, 1976) (Miles and Philbrick, 1988) (Arendt, 1987). Experimental administration of melatonin has been shown to cause a more rapid induction of sleep and increase the depth and intensity of sleep resulting in a more restful feeling upon awakening (Waldhauser et al, 1984). The effects of melatonin on sleep have led to its use as a folk cure sleep tonic among many people worldwide.

Cancer

Malignancies occur at all ages and at all stages of human development, but are known to steadily become more common with advancing age. This correlation of cancer rates with age have led many to wonder whether cancer is in some way directly related to the aging process. Perhaps the mechanisms that causes cancer also cause aging, or vice-versa.

Conversely, scientists have also conjectured that the transformation of a normal cell into a malignant cell is a multi-step process that takes a long time to occur, thus yielding the relationship between aging and cancer. Clearly there is no present answer to this question of whether cancer is related to aging. This may be true because both cancer and aging may have related root causes, or simply that it takes time for cancer to develop and for the aging process to occur. The relationship of melatonin with cancer may help to shed light on these questions.

Melatonin administration has indeed been found to inhibit the growth of several types of malignancies. The growth of breast tumors, leukemia, and prostate cancers are inhibited by melatonin administration (Lapin, 1974) (el-Domeiri et al., 1973) (Toma, 1987) (Buswell, 1975) (Karmali et al., 1978). The treatment of sarcomas has been improved with melatonin (Star, 1970). Patients suffering from several types of malignancies including breast, lung, stomach, bone osteogenic sarcoma, acute leukemia, chronic leukemia, and lymphomas survived longer and reported reductions in symptoms when treated with melatonin. Other patients suffering from tumors which had metastasized reportedly improved, even though they had previously failed to respond to conventional therapies (Lissoni, 1989).

Several interesting correlations of the pineal gland and cancers have been observed. Changing the endocrine control conditions of the body by excision of the pineal gland, which produces melatonin and other control chemicals, seems to influence tumor growth. Surgical removal of the pineal has been shown to stimulate the growth of breast tumors, ovarian cancers, sarcomas and fibrosarcomas, melanomas of hamsters, and Ehrlich's tumors (Lapin, 1976) (Lapin, 1974), (Barone, 1972), (Rodin, 1973), (DasGupta, 1967), (Tamarkin, 1981).

On the other side of the equation, it has been discovered that cancer patients have larger pineal glands than did other persons studied (Rodin, 1967). This correlation of larger pineal glands in cancer patients along with the findings that removal of the gland leads to enhanced tumor growth, causes us to wonder if the pineal produces substances which help to inhibit cancer formation or growth. If so, then perhaps the gland becomes larger in cancer patients as it too produces as much of the anti-tumor substance as possible. Could melatonin be the anti-tumor substance being produced?

In experiments in which the pineal had been removed, melatonin inhibited the growth of both melanomas and Yoshida tumors. However, these findings in melanomas and Yoshida tumors did not occur when the pineal remained intact, possibly indicating that the anti-cancer effect of endogenously produced melatonin was not significantly augmented by the additional melatonin.

How does melatonin produce these salutary effects on cancerous growths? Three theories have been promulgated. Several authors have suggested that melatonin inhibits malignancies by disrupting the production by the body of other hormones which, themselves, promote tumor development or growth (Star, 1970) (DiBella, 1979), (Blask and Hill, 1988).

Melatonin has also been suggested to directly effect cancer cells by blocking the cancer cell's ability to be stimulated by other hormones found in the body. An interesting example of this phenomenon is melatonin's possible role as a natural estrogen blocking agent (Blask and Hill, 1988). In this regard, melatonin seems to act like the exciting new breast cancer drug Tamoxifen. Lastly, melatonin may work via stimulation of the immune system. The immune cells are thought to play a role in surveillance of the formation of new cancer cells and then in the cancer cells' destruction after detection.

Role In Metabolism

Caloric intake by dietary restriction has long been acknowledged as a factor predisposing to longevity. Experiments in animals revealed near-starvation to cause increased life expectancies when other disease processes were eliminated from the analysis (Segall, 1979). Low calorie, nutritionally supported dietary regimens have also shown significant effects including prolongation of life and delayed sexual maturation in animals (Segall, 1979). Studies of survivors of long periods of dietary deprivation in war prisoner camps and holocaust victims showed increased life spans when diseases are factored out.

The body seems to be propelled during active, daylight hours by serotonin, which enhances the anabolic growth and reproduction functions of the lifecycle. Melatonin appears to act as a counter-balance during the resting, nighttime portion of the lifecycle to promote repair and rejuvenation.

Melatonin can be suppressed by chronic lack of sunlight exposure on the retinas, as well as through underfeeding. One way to chronically and permanently stop sunlight exposure on the retinas of experimental animals is to surgically blind them. This blinding serves to totally stop any and all sunlight exposure reaching the retina. The pineal gland then increases its production of melatonin since, similar to a bear's melatonin-based hibernation response in the dark northern winter, the brain perceives that it is in endless night.

Studies have shown that such experimental blinding causes changes in several hormone levels. These changes include increased melatonin production, which then decreased the levels of serotonin, prolactin, growth hormone, leuteinizing hormone, and follicle stimulating hormone (Blask, 1984). This group of five hormones, which are decreased by melatonin, serves the organism during growth, reproduction and heavy metabolic activity. They may therefore support the most strenuous activities of life. These strenuous activities may possibly produce waste by-products, cellular damage, or other unknown sequelae that may require rest and melatonin to repair. An important role for "sleep" in mammals may therefore be repair and rejuvenation made possible by a preponderance of central nervous system and systemic melatonin at night or a higher melatonin/serotonin ratio during sleep.

Animal dietary deficiencies in the amino acid tryptophan, a metabolic precursor to serotonin, causes symptoms of melatonin enhancement similar to those seen in blinding and generalized caloric restriction (Segall and Timiras, 1975). Researchers found these experimental animals to have delayed sexual maturation and growth, decreased incidences of cancer, and added longevity. Additionally, the deficiency of tryptophan decreased the animals' body temperatures and metabolic rates, similar to the effects of melatonin augmentation. These effects may be caused by a relative paucity of serotonin in relation to melatonin and a resultant enhanced melatonin/serotonin ratio.

Summary Of Melatonin And Aging

Central nervous system production of melatonin may be determined by a cellular time clock within the pineal gland cells. It is based upon the gradual deterioration of the telomeres of the pineal cells. As the telomere time clock of the pineal gland cells ticks away, it modifies the rate of secretion into the general bloodstream of melatonin and its precursor, serotonin. The gradual changes in melatonin and serotonin may then serve to set our life-clock and determine our longevity. Melatonin, the predominant nighttime hormone, may serve to counteract the sequelae of serotonin, the predominant daytime hormone. A greater amount of serotonin during the active, daylight hours may serve to enhance increased metabolism, growth, maturation, and reproductive activities.

A greater prevalence of melatonin during the restful nighttime hours may produce repair and rejuvenation of the deleterious sequelae of daytime serotonin including waste byproducts and cellular damage. Nighttime melatonin may also temporarily set our metabolic "thermostat" at a lower level. The resultant effects of melatonin's rejuvenating properties may add up to longer life spans and lessened rates of malignancies.

A gradual breakdown of this restorative system following early adulthood may lead to the eventual signs of senescent deterioration. The concept of "sleep therapy" can also include the use of GABA and GHB. Both of these substances are being studied in leading research universities for their roles in promoting health changes during sleep. They may act to restore normal youthful brain functioning and help to reset hormonal levels to that found in younger persons.

How To Use Sleep Therapy

Measuring the middle-of-the night melatonin level gives us the first clue as to our current state. We normally want to measure the level at about 2:00 am, so that we have people set their alarm, wake up briefly to obtain a saliva sample, and then it is mailed to the lab the next morning. If the level is below normal for the age of the patient, or if it seems wise to try to replace melatonin to a level of that of a younger individual, we may consider time-released melatonin supplementation. This type of melatonin helps to more closely mimic the natural nighttime pattern of melatonin production seen in young individuals. BioMelatonin contains 2mg of timed released melatonin per tablet. I often prescribe one to three tablets to be taken an hour prior to bedtime. The dose depends upon the saliva test results and the response to varying doses.

Case History:

A patient of mine complained of sleepless nights, which she called insomnia. Careful questioning made anxiety syndrome and clinical depression unlikely. A 2:00 am saliva test to determine her melatonin level revealed it to be quite low. Guessing that she may require 4 to 6 mgs of melatonin to correct the deficit, I nonetheless began dosing at just 2mg each bedtime. I did this in order to allow her physiological responses to guide us in coming to the final dose. Two milligrams had no effect, but adding another 2 mg tablet at bedtime gave her a sound and restful sleep. As do many patients, she says that her entire quality of life has been improved now that she rests soundly each night.

References

Klein, D. C. "The Pineal Gland: A Model Of Neuroendocrine Regulation," *Res Public Association Res Nerv Mental Disease* 56 (1978): 303-27.

Klein, D. C., Weller, J. L., "Indole Metabolism In The Pineal Gland: A Circadian Rhythm In N-Acetyltransferase," *Science* 169 (1970): 1093-95.

Welsh, M. G., "Pineal Calcification: Structural And Functional Aspects," *Pineal Res* Rev 3 (1985): 41-68.

Rozencweig, R., Grad, B. R., Ochoa, J., "The Role Of Melatonin And Serotonin In Aging," *Medical Hypotheses* 23 (1987): 337-52.

Miriam, M., Schwaab, D. F., Witing, W., Honnebier, M., van Gool, W. A., Eikelenbloom, P., "Biological Clocks In Aging, Development, And Alzheimer's Disease," *Brain Dysfunction* 2 (1989): 57-66.

Iguchi, H., Kato, H., Ibayashi, H. "Age Dependent Reduction in Serum Melatonin Concentration in Healthy Human Subjects," Journal of Clinical Endocrinology 55 (1982): 27-29.

Pierpaoli, W., Maestroni, J. M., "Melatonin: A Principle Neuroimmunoregulatory And Anti-Stress Hormone: Its Anti-Aging Effects," *Immuno Lett* 16 (1987): 355-62.

Pierpaoli, W., Dallara, A., Penrinis, E., Regelson, W., "The Pineal Control Of Antiaging: The Effects Of Melatonin And Pineal Grafting On The Survival Of Older Mice," *Ann NY Acad Science* 621 (1991): 291-313.

Anisomov, V. N., Bonderenko, L. A., Khavinson, V. K., Morozov, V. G, "The Pineal Peptides: Interaction With Indoles And The Role In Aging And Cancer," *Neuroendocrinology Lett* 11 (1989): 235.

Pierpaoli, W., Yi, C., "The Involvement Of Pineal Gland And Melatonin In Immunity And Aging, 1. Thymus-Mediated Immunreconstituting And Anti-Viral Activity Of Thyrotropin Releasing Hormone," *Journal Neuroimmunology*, 27 (1990): 99-109.

Wurtman, R. J. Altshule, M. D., Holmgren, U., "Effects Of Pinealectomy And Of A Bovine Pineal Extract In Rats," *American Journal of Physiology,* 197 (1959): 108-10.

Gromova, E. A., Kraus, M., Krecek, J., "Effect Of Melatonin And 5-Hydroxytryptamine On Alsosterone And Corticosterone Production By The Adrenal Glands Of Normal And Hypophysectomized Rats," *Journal of Endocrinology*, 39 (1967): 345-50.

Jouan, P., Samperez, "Rechurches Sur La Specificite De Action De La 5-Hydroxytryptamine Visavis De La Secretion In Vitro De D'aldosterone," *Ann Endocrinology* (Paris) 25 (1964): 70-75.

Trentini, G. P., Genazzani, A. R., Criscuolo, M., Petraglia, F., DeGaetani, C., Ficarra, G., Bidzinska, B., Migalsi, M., Genazzani, A. D., "Melatonin Treatment Delays Reproductive Aging Of Female Rats Via The Opiatergic System," *Neuroendocrinology* 56 (1992): 364-70.

Smythe, G. A., Lazarus, L., "Growth Hormone Responses To Melatonin In Man," *Science* 184 (1974): 1373-74.

Maestroni, G. J. M., Conti, A., Pierpaoli, W., "Role Of The Pineal Gland In Immunity: Circadian Synthesis And Release Of Melatonin Modulates The Antibody Response And Antagonizes The Immunosuppressive Effects Of Corticosterone," *Journal Neuroimmunology* 13 (1986): 19-30.

Lissoni, P., Tancini, G., Rovelli, F., Cattaneo, G., Archili, C., Barni, S., "Serum Interleukin-2 Levels In Relation To The Neuroendocrine Status In Cancer Patients," *European Journal of Cancer* 62 (1990): 838-39.

Bailowsky, S., "Effects Of Melatonin On The Photosensitive Epilepsy Of The Babboon, Papio Papio," *Electroencephal Clin Neurophysiology* 41 (1976): 314-19.

Miles, A., Philbrick, D. R. S., "Melatonin And Psychiatry," *Biol Psychiatry* 23 (1988): 405-25.

Arendt, J., Aldhouse, M., Marks, M., Folkhard, S., English, J., Marks, V., Arendt, J. H., "Some Effects Of Jetlag And Its Treatment By Melatonin," *Ergonomics* 30 (1987): 1379-91.

Waldhauser, F., Lynch, H. J., Wurtman, R. J. "Melatonin In Human Body Fluids: Clinical Significance," *The Pineal Gland*, ed. R. J. Reiter, Raven Press, New York (1984) 345-70.

Lapin, V., "Influence Of Simultaneous Pinealectomy And Thymectomy On The Growth And Formation Of Metastases Of The Yoshida Sarcoma In Rats," *Exp Path* 9 (1974): 108-110.

El-Domeiri, A., Das Gupta, T. K., "Reversal By Melatonin Of The Effect Of Pinealectomy On Tumor Growth," *Cancer Research* 33 (1973): 2830-33.

Toma, J. G., Amerongen, H. M., Hennes, S., O'Brien, M. G., McBlain, W. A., Buzzell, G. R., "The Effects Of Olfactory Bulbectomy, Melatonin And/Or Pinealectomy On Three Sublines Of The Dunning LR3327 Rat Prostatic Carcinoma," *Journal Pineal Research* 4 (1987): 321-39.

Buswell, R. S., "The Pineal And Neoplasia," *Lancet* 1 (1875): 34-35.

Karmali, R. A., Horrobin, D. F., Ghuyur, T., "Role Of Thepineal Gland In The Etiology And Treatment Of Breast Cancer," *Lancet* 2 (1978): 1002.

Starr, K. W., "Growth And New Growth: Environmental Carcinogens In The Process Of Human Ontogeny," *Prog Clin Cancer* 4 (1970): 1-29.

Lissoni, P., Barini, S., Crispino, S., Tancini, G., Franschini, F., "Endocrine And Immune Effects Of Melatonin Therapy In Metastatic Cancer Patients," *European Journal of Cancer Clin Oncol* 25 (1989): 789-95.

Lapin, V., "Pineal Gland And Malignancy," *Osterreich Zeitsch Onkol* 3 (1976): 51-60.

Barone, R. M., Abe, Rl, Das Gupta, T. K., "Pineal Ablation In Methylcholanthrene Induced Fibrosarcomas," *Surgical Forum*, 23 (1972): 115-16.

Rodin, A. E., "The Growth Spread Of Walker 256 Carcinomas In Pinealectomized Rats," *Cancer Res* 23 (1963): 1545-50.

Das Gupta, T. K., Terz, J., "Influence Of Pineal Gland On The Growth And Spread Of Melanoma In Hamsters," *Cancer Research* 27 (1967): 1306-11.

Tamarkin, L., Cohen, M., Rosselle, D., "Melatonin Inhibition And Pinealectomy Enhancement Of 7,12-Dimethylbenzenethracene-Induced Mammary Tumor In The Rat," *Cancer Research* 41 (1981): 432-36.

Rodin, A. E., Overall, J., "Statistical Relationships Of Weight Of The Human Pineal To Age And Malignancy," *Cancer* 20 (1967): 1203-14.

DiBella, L., Rossi, M. T., Scalera, G., "Perspectives In Pineal Functions," *Prog Brain Research* 52 (1979): 475-78.

DeBlask, S. M. Hill, "Melatonin And Cancer: Basic And Clinical Aspects," *Melatonin: Clinical Aspects* Oxford University Press (1988): 128-73.

Chapter Eleven
PSYCHOLOGY OF THE MIND-BODY CONNECTION

Psychological factors can have a great influence on our state of health, and upon the functioning of our physical bodies. Our spiritual, mental, and psychological health can all affect the healthy functioning of our physical being. Indeed, this interconnection between mind, body and spirit is slowly and grudgingly being recognized by conventional medicine. Meditation, personally preferred religious practices, and psychological counseling are slowly gaining acceptance as important components of healthcare. The University of Arizona's Doctor Andrew Weil, M.D., and Deepak Chopra, M.D., have been bringing this long-neglected realm of health to the forefront.

The Mind-Body-Spirit Connection in Meditation for Health

Meditation, prayer, and directed visualization are all means by which we can attempt to focus positive, nurturing, healing energy on our physical body. These practices engender the most common ways in which humans attune the mind and spirit to the ailing physical body. Focusing nurturing thoughts via meditative techniques is thought to allow the brain to influence the rest of the body in positive ways. This may occur through the release of healing hormones, via direct nerve stimulation of distant ailing areas, or via the brain's influencing of "the vital force," a concept long revered in homeopathy and many ancient healing traditions.

The concept of "vital force" means that our physical body may be animated by a force that flows throughout the physical person dictating the state of health or disease. A disrupted vital force may allow for disease to manifest, whereas correcting a deranged vital force may lead to the healing of disease states. It is postulated that our brains, via meditative processes, may be able to rectify a disordered and therefore disfunctional vital force. Other modalities that are thought to correct an abnormal vital force include acupuncture, in which needles physically stimulate abnormal channels of vital force, and classical homeopathy, in which physical substances are "potentized" to produce a medicine that acts by correcting an abnormal vital force. Homeopathy utilizes the concept of "like treats like." This means that a substance that can cause illness can be processed in such a way that it actually cures the same symptoms of illness, instead of causing them.

Meditiation and prayer seek to connect the mind, the spirit and the body together in a more profound way than usually occurs during our day-to-day activities. When we focus on a higher order, we may drown out, temporarily at least, the common thoughts and activities of daily living that distract us from self-healing.

There are many ways in which we may choose to accomplish this, but a certain technique known as directed visualization is gaining popularity. Directed visualization occurs as we take a quiet time to focus our mind on healing thoughts. Meditating on thoughts of the body strengthening itself, on the immune system reacting against the disease, and on diseased cells "self-destructing" are becoming very popular.

Is it truly possible that our mind can profoundly affect our bodily functions? More and more evidence is mounting that links chronic negative mental states to various diseases. Certain personality types are said to be at greater risk of developing heart disease or cancer. Additionally, evidence is mounting that our spiritual and mental states can effect our ability to perform seemingly unbelievable tasks.

Religious zeal is sometimes manifested as a psychological state known as "religious ecstasy." Persons in the ecstatic state may see visions, hear the voice of spiritual beings, and even manifest super-human feats of strength or endurance (as in firewalkers). Let's next examine one particularly unusual physical manifestation, which can occur during religious ecstasy. This is known as "speaking in tongues." It serves as an example of just how profoundly the mind, body and spirit are interconnected, and the significance of the spiritual being.

Religious Ecstasy: The Mind-Body-Spirit Connection

Intensely emotional and, to many, shockingly exuberant manifestations of worship attributed to the Holy Spirit that accompanied Pentecostal worship services, have perplexed on-lookers since the days of the book of Acts. These astoundingly exuberant forms of worship, ranging from speaking in tongues to dancing in the spirit and prophecy are unusual to most other Christian groups.

Heartfelt worship and acts of faith accompanied by supernatural gifts have for centuries been anathema to contemporary religion, whether it be the Judaism of *New Testament* times, Roman Catholicism of the Middle Ages, or present day reformation religions and humanistic philosophy. As with any other group who are in some substantial way different from the cultural moré, Pentecostal groups have caused much conjecture as to the nature of their unusual form of worship, and to what factors make these people so different from the religious norm. At different times in history, evaluation by secular religious leaders led them to the conclusion that Pentecostalism was inspired by evil (demons, witchcraft, or rebelliousness), and that pentecostals acted against the established religions' authorities.

Manifestations of the misconceptions about such worship have led to unfortunate incidents of cruelty. Many sources from ancient times and the medieval period attest to these instances. From the Christian *New Testament,* Book of Acts, in the second chapter come the following verses:

> And when the day of Pentecost was fully come, they were all with one accord and one place. And suddenly there came a sound from heaven as of a rushing mighty wind, and it filled the entire house where they were sitting. And there appeared unto them cloven tongues like as of a fire, and it sat upon each of them. And they were all filled with the Holy Ghost, and began to speak with other tongues, as the spirit gave them utterance. Now when this was noised abroad, the multitude came together, and were confounded, because every man heard them speak in his own language. And they were all amazed, and marveled, saying one to another, "Behold, are all not these which speak Galileans? And how hear we every man in our own tongue, where in we were born? We do hear them speak in our tongues the wonderful works of God." And they were

all amazed, and were in doubt, saying one to the other, "What meaneth this?" Others mocking said, "These men are full of new wine." But Peter standing with the eleven said, "These are not drunken, as ye suppose. But this is that which was spoken by the prophet Joel; and it shall come to pass in the last days, sayeth God I will pour out my spirit upon all flesh, and your sons and your daughters shall prophecy and your young men shall see visions, and your old men shall dream dreams."

Acts 2:1-17

Again, in the Book of Acts, Chapter 4:

And when they had prayed, the place was shaken where they were assembled together and they were all filled with the Holy Ghost, and they spoke the word of God."

Acts 4:31

We see evidence throughout the Book of Acts that the boldly, powerful manifestations of worship caused much persecution of the early believers.

In Acts 5:17: Then the high priest rose up and all they that were with him, and were filled with indignation, and laid their hands on the apostles, and put them in the common prison."

Acts 5:28 reads, "As the apostles were brought before the Sanhedrin saying, 'Did not we straightly command you that ye should not teach in this name? And, behold, ye have filled Jerusalem with your doctrine, and intend to bring this man's blood upon us.'"

Acts 5:40 reads, "And when they had called the apostles, and beaten them, they commanded that they should not speak in the name of Jesus and let them go."

The Book of Acts of the Apostles is replete with the stories of the persecution of these early believers. From Acts 8:1; 22:4; and 26:11, we find these early believers persecuted in every city and area in which they were deemed unusual.

During the Medieval period of history, practitioners of these unusual forms of worship were persecuted more cruelly than ever before. Christians of various other beliefs seemed to find the Pentecostal Christians even more of a danger to their way of life than the Church in the Book of Acts was found to be by the Jews, Romans and Greeks of their times. (Note: reference Fox's *Book of Martyrs* for more material.

As Renaissance scientific thought gained preeminence over religious explanations, philosophers and psychologists began to develop theories based on scientific thought to explain Pentecostalism. These theories were, for the most part, based on the assumption that Pentecostals were mentally ill or psychologically impaired. The manifestations accompanying their worship was surmised by some to be a direct manifestation of psychopathology and by others to be a therapeutic action with which to resolve and destroy the inner psychological conflicts of the worshipper. Until recently, these theories based on psychological pathology, were preeminent. Research, however, is tending to refute the claim of psychological illness being the basis of Pentecost.

Several of the much acclaimed, long-standing explanations of Pentecostal phenomena are based upon the assumption that tongues speakers enter easily into emotionally hyperexcited states. The state of hysteria has commonly been employed to describe the basis of the Pentecostal experience.

Hysteria is defined medically as a neurosis in which involuntary psychogenic loss or disorder of function occurs. The conversion type of hysterical neurosis involves the loss of voluntary motor or sensory functions, such as feelings of anesthesia or paralysis. It can also include sensory hyperacuity or motor hyperacuity including ticks, tremors, or seizure-like epileptic appearing activities. This disassociative type of hysterical neurosis is described by characteristic alterations in consciousness of personal identity and can include such symptoms as amnesia, fatigue, sleepwalking (also known as somnambulism), or multiple personality characteristics. Interestingly, many of the disassociative disorders have an abrupt onset commonly following periods of emotional stress as opposed to other illnesses that occur when some physical or biological insult occurs to the brain.

In contrast to the dissociative disorders, conversion disorders are characterized more by psychogenic loss and/or disturbances of bodily functions, rather than being alterations on the person's state of consciousness or state of his own personal identity. Some of the most commonly observed symptoms of conversion seen by a psychiatrist include altered sensation states. These states of altered sensation can occur as anesthesia or parathesia, with anesthesia being a broadly felt lack of physical sensation of the body and parathesia being more transient and localized to specific areas of the body. Partial or complete blindness or deafness has also been seen in these patients. A second form of conversion disorder involves paralysis and/or the inability to stand or walk correctly. Ataxia can also occur with paralysis and may be manifested by the loss of the ability to speak, commonly known as aphonia, and even the ability to stand from a sitting position.

Another type of conversion disorder involves disconesias or involuntary movements. Within this group of phenomena include the pseudo-epileptic seizure activities seen in some patients. Still another form of conversion disorder involves the sensation of pain that can effect the various parts of the body such as headaches, pains in the chest, and abdominal cramping among many others. This last type of conversion disorder involving pain is thought by psychiatrists to be the most common form of the conversion hysteria.

As in dissociative disorders, the onset of the symptoms of conversion disorder is many times sudden and dramatic. Scientists believe that the psychiatric patient obtains psychological benefit from the appearance of these symptoms of conversion disorder. Their physically based disability can result in what is known as primary gain, "a feeling of relief from emotional conflict and anxiety," and what is known as secondary gain, "the driving of external benefits such as being released from responsibilities or the achievement of increased personal attention or financial benefits."

Cutten, in 1927, summarizes several aspects of hysteria which his contemporaries believed tied it to tongue speaking and religious ecstasy. He wrote that hysteria commonly causes normal sensations to be exaggerated or reinterpreted by the mind as something different from reality, that the hysteric normally remains fully conscious, that normal self-control is not evident, and that hysterics are more susceptible to suggestions than are normal people.

Cutten also summarized the thoughts that the ecstatic state or catalepsy were the source for glossolalia ecstasy, an emotionally exalted state in which the individual loses much of his self-control and powers of rational thought. The person is left less self-conscious and is out of touch with external events while in this state.

Some believe that people can learn to voluntarily enter the state or that it may be induced through the effect of external stimuli such as an emotion-charged revival or other religious meetings. Aspects of ecstasy thought to apply to tongues are that in ecstasy, thought is commonly focused on one dominant idea. Varying degrees of consciousness may be seen and in cases of unconsciousness, the person may later recall having visual or auditory experiences during the ecstatic loss of consciousness. Ecstatics are also thought to be emotionally liable and to be prone to express their emotional turmoil through speech and unusual body movements.

Others believe that a state known as catalepsy plays a role in speaking in tongues. Medically, this is also known as cerea flexibilitas or waxy flexibility. Catalepsy is seen in certain condition such as catatonic schizophrenia. In this condition, a person exhibits wax-like yielding and can be placed in any condition. The position will be maintained for periods as long as several hours.

The proponents of the catalepsy theory do not believe that complete catalepsy is seen in Pentecostalism, but do think that a partially developed cataleptic state may exist. They argue that changes in muscle tone, suspension of sensation, loss of consciousness, loss of volitional control, nervous instability, impassioned exclamations uttered while unconscious, and visions or hallucinations occurring during unconsciousness are all aspects of catalepsy which may form the basis for speaking in tongues and related experiences.

Schizophrenia, in general, has also been implicated by some scholars as the source of the unusual manifestation of the Pentecostal experience. They feel that tongue speakers are possessed by the contents of their unconscious. These scholars are evidently attempting to link speaking in tongues with the speech disorder seen in many schizophrenics.

Schizophrenia is a psychotic mental disorder of thinking, mood and behavior. Several types of specific speech anomalies are seen with schizophrenia. Schizophrenics are often incoherent, uttering meaningless sentences due to their impaired mental ability to associate their ideas in a realistic manner. They are also known to use neologisms, which are meaningless, completely new words coined by the schizophrenic speaker to express a thought or idea. Echolalia, the meaningless repetition of sentences or words heard by the schizophrenic, is also often exhibited.

Another anomaly of speech seen in schizophrenia is verbigeration, a senseless repetition of words or phrases that may last for days at a time. As can be seen, schizophrenics may exhibit any one of these several types of unusual speech patterns.

Certain behavioral and emotional disorders of schizophrenia may also be confused with the kinetic and supernatural components of Pentecostal worship. Stereotyped behavior consisting of repetitive performance of unusual gestures of repetitive patterns of walking or moving is seen in certain schizophrenics. They sometimes also exhibit the imitation of gestures and movements seen in others, a symptom known as Echopraxia, thought to be analogous to Echolalia, the verbal repetition symptom.

Delusions, sometimes with religious implications, and abnormal states of inappropriate emotions are seen in schizophrenics. Delusions of grandeur or of being controlled by an external force such as gods or space aliens have been reported. Schizophrenics may, at times, exhibit states of emotional exaltation with grandeur feelings of openness with the universe, anxiety over thoughts of an impending end of the world, religious ecstasies, or feelings of omnipotence.

Some authors feel that a result of delusional or other abnormal mental states may cause a heightened power of memory to occur. It is suggested by some writers that this exalted memory may form the basis for speaking in tongues and they feel that this may be particularly applicable to the occasions on which it is claimed that understandable foreign languages are heard during speaking in tongues. The proposed scenario is that sometime in the tongue speakers' past they had some experience with a foreign language. When the person later entered into a delusional state, which the theorists argue occurs during a Pentecostal worship, an unusual power of memory occurs enabling a speaker to produce words or phrases of the language heard years previously. Several examples in which this may have happened, although not involving Pentecostal speaking, are given by Cutten.

Hypnotism has long been argued to be an important factor in the production of speaking in tongues. Mosiman, in an important early work on tongues, *Das Zungenreden*, stated "The ecstatic condition is really a hypnotic one." But is this manifestation really due to a hypnotic trance effect, or a more profound interaction of the spirit, mind and body?

Hypnosis is a state resembling sleep but which is brought on by another person known as the hypnotist whose suggestions are accepted readily by the person being hypnotized. Hypnosis is therefore an induced sleep-like, trance-like state well known in both science and medicine. It is surmised that the hypnotic state allows the mind a greater influence over the body, perhaps for healing, and maybe to induce the ability to speak in tongues.

Several factors, which are conditions to induce the hypnotic state, are argued as being present in Pentecostal meetings. These factors are described by Cutten as "fixations of attention, uniformity of perception, limitation of the power of will, limitation of the field of consciousness, impression of ideas."

The listing of components of Pentecostal services is typical of the opinions of many of the researchers who have observed such worship. A rapt attention given to the preacher or that focused upon God at various times during a common Pentecostal service, and the usual desire of the worshippers to "yield one's self to the Spirit" are the basis of many scholars' theories that a mass hypnotism or perhaps auto hypnotism is occurring.

An important factor in the inducement of hypnotic states is that of the level of the suggestibility of the subject. A person with increased suggestibility is thought to be a good subject for hypnotism. It has been suggested that in the process of "yielding one's self to the Spirit," a Pentecostal worshipper is actually increasing his level of suggestibility. It is surmised that when this occurs, a hypnotic state is produced in which higher powers of conscious control are decreased and lower, more reflexive levels of consciousness gain pre-eminence in producing the subject's thoughts and actions. In this condition, the preacher is thought to have great influence and can cause the worshipper, either through direct verbal suggestion or implied suggestion, to produce unusual physical manifestations such as speaking in tongues, prophecy, and dancing in the spirit.

The practices of fasting and long prayer vigils are also argued to aid the production of the hypnotic state. Mosiman sums up this theory of how tongues are produced: "The gift of tongues is the expression of thought and feeling by the speech organs, which temporarily come under the control of the reflex nerve centers, the peculiar forms of which are determined by suggestion which chiefly consists of a verbal interpretation of the New Testament."

The way in which Pentecostal services are conducted is not the only factor which researchers have implicated as heightening suggestibility. It has been suggested that Pentecostals are, as a group, inherently more suggestible than other people. It has also been surmised that glossolalists are generally less educated and less articulate. It is thus argued that because they have an impaired vocabulary and decreased verbal ability, they find it difficult to express their deeply ecstatic religious feelings during worship. It is suggested that they therefore use infantile type babble, or words of their own coining to vocally express their feelings, and thus speaking in tongues is produced.

The following course of events is thought to occur: 1) a literally impaired worshipper finds himself in church with its sources of suggestion and in an excited condition, 2) the state of excitement drives him to speak, 3) although the speech may begin normally, the combination of nervous energy and subnormal powers of verbal expression cloud the mind and he is unable to express his thoughts vocally, 4) higher mental functioning loses its pre-eminence and lower centers of thought and functioning gain control, 5) a trance-like state develops, and since the suggestion is to speak, meaningless syllables and phrases are spoken, and 6) the result in gibberish is unintelligible and so assumed by other worshippers to be a foreign language.

Some authors also suggest that not only are tongue-speakers more suggestible, but they are also weaker and more dependent psychologically. It is argued that because of this personality type, they seek and are more prone to dogmatic rules of group conduct.

A very intriguing theory on the production of tongues is that of automatism. The concept of automatism was developed by Bernheim in his research into different types of hysteria. An automatism is an act that the human mind or body produces while under the influence of the unconscious state. These acts are not under conscious or volitional control.

Automatism can be either of the sensory or motor type. Sensory automatism includes visions, auditions, and other hallucinations of perception, which all occur with such intensity that the person sensing them is sure that they are real. Motor automatism is seen when physical action is performed, such as automatic writing, unconscious speech or "phonetic automatism" and movements of the arms or walking. It has been suggested that during worship a type of unconsciousness may ensue resulting in sensory and/or motor automatism. This theory has been in existence for many decades and is similar to the theory of the linguistic researcher F. D. Goodman.

Lombard in "De la Glossalalie" proposes that tongues may result when powerful emotional tendencies are brought to the surface during worship services. He thought that the tendencies of the personality, known as complexes, are normally repressed by the conscious mind and thereby hidden and detained deep within the psyche. He suggested that when these complexes are allowed to surface during emotion-laden worship, they help to cause the formation of automatism.

There are many theories brought forth to explain the linguistic aspects of speaking in tongues. The Pentecostals are convinced that tongues are a God-given prayer in a distinct and linguistically correct language, whether it be in a primary earthly language, a certain dialect, or an unknown "angelic tongue."

Dr. W. Samarin is a prominent professor and linguistic scholar who has studied the phenomenon on tongues from a scientific perspective. From his research, Dr. Samarin has developed his own theory of the nature of glossolalia. In scientific circles, his theory is known as the Sociolinguistic theory of Glossolalia. Perhaps by developing an understanding of Dr. Samarin's research findings and the conclusions he draws from these findings to form his sociolinguistic theory, we can gain some added insight into what is expressed by Pentecostals through speaking in tongues. They may also gain some insight into how researchers of linguistic analysis are influenced by their own cultural and religious backgrounds and by their preconceived notions of the theological basis of the groups they are studying.

The central thesis of the sociolinguistic theory is that people either learn or teach themselves to produce a language-like group of sounds. The speakers thought to be motivated by pressure from the religious group or by expectations derived from their own religious beliefs to produce an "unknown foreign prayer language."

Let's begin our discussion on Dr. Samarin's work by exploring his thoughts on the linguistic nature of tongues. Basically, Samarin believes that glossolalia is a deviant or anomalous state. In that, as he sees it, tongues are meaningless and follow no set rule of grammar.

From his analysis of taped and written passages of speaking in tongues, several conclusions as to its linguistic nature have been made. Glossolalic discourse is thought to be commonly divided into "sentences" or perhaps more precisely, "breath groups." Although ubiquitous, the nature, length of sentences, and length of pauses between sentences, is seen to vary from individual to individual. Sometimes it is noticed that certain people will speak with a repetitious "melody-like sound" in a repeated pattern. It is noted, though, that even in English, certain individuals have their own unique pattern of intonation, patterns or sentence melody when in prayer.

An attempt has been made to analyze glossolalia on the basis of the relationship of syllables to each other in an utterance. This is known as the syntagmatic relationship of syllables. For example, one might wish to ascertain if a pattern exists such as, with the syllable "lo." To do this, a sample of speech would be analyzed to determine if this syllable "lo" has any certain pattern, such as if it were to always precede another certain syllable or if it were found to always begin a sentence or always end a sentence.

In studying discourses from this perspective, Samarin was unable to make any conclusions as to whether actual words or sentences are being produced by tongue speakers. He does note that certain speakers tend to repeat certain sounds or phrases, "words with a question mark," more commonly than other people do. The difficulty in assigning the function of words or sentences to different groups of sounds in a glossolalic example is that the researcher has no clue as to the meaning or grammar structure of the language, if it is indeed a language being spoken.

When certain texts were analyzed by a group of linguists, more of them gave the same interpretation concerning what they found to be present in the sample of glossolalic speech. Samarin believes that this impairment in the ability of linguists to analyze spoken glossolalic stems from the fact that there is no given meaning for the utterance being studied. He contends that these scientists are unable to accurately ascribe words or sentences when studying spoken narrative passages if the scientists are unable to discern any apparent meaning from the utterances they are studying.

So then, if analysts such as Dr. Samarin try to draw a conclusion based on examining glossolalia as words and sentences, their conclusions should be very suspect because the linguistic scientists themselves admit that it is difficult, if not impossible, to analyze glossolalia by using their normal means of linguistic analysis.

In our further study we will see how another technique of linguistic analysis can be employed to elucidate the phenomenon of speaking in tongues. While examining Dr. Samarin's findings and conclusions, we must keep in mind that Dr. Samarin has admitted that there are important inherent limitations in his method of analysis.

Dr. Samarin believes that one of the most basic and important aspects in what he describes as the "production" (i.e., the sociolinguistic theory) of glossolalia is repetition. This repetition involves the repeating of syllables, consonants, vowels and even entire phrases. He feels that the most important factor is the repetition of consonants. For example, if a certain consonantal word skeleton is employed repeatedly but with different vowels each time, the consonantal word skeleton could sound as if several different words were being spoken. Complete repetition of identical elements, however, was not found commonly in the samples Dr. Samarin studied.

If speaking in tongues is thought of as being an actual language with words and phrases in the actual earthly language being spoken, then certainly the meaning in the language being spoken in the tongues should be apparent to anyone with knowledge about that language. Therefore, when scientists argue that no understandable meaning is present in a specimen of tongues speaking, they are assuming that if they are unable to recognize the language being spoken, then it is not a real language.

However, there are many known primary languages and dialects present in the world today. In addition to these currently spoken languages, an unknown number of ancient and now extinct languages have existed in the past. Is it also possible that some languages spoken during these times are not earthly in origin at all but are, as the Bible says, angelic or heavenly in nature? Is it perhaps possible that an understandable meaning may indeed be present, but not to the earthly listeners such as scientists who are studying the tongue-speakers?

This should not surprise us, however, if tongues are understood essentially as an expression of the spirit during its prayer to the Almighty with the meaning perhaps in some cases best understood only in the communion between man and creator. It should also be noted that many charismatic ministers claim that tongues does not indeed need to be a real language either earthly or angelic, but may simply be an audible expression of ecstatic feelings during prayer.

Dr. Samarin has noticed the similarity to prayer names or proper names from the Bible and words spoken commonly by certain glossolalists. He cites having heard Jezu or Jeshua, which he thought was referring to Jesus. It should be noted, incidentally, that Jeshua is thought to be the original Hebrew pronunciation of what we now pronounce as the Greek word, Jesus.

Dr. Samarin also reports his findings in which he believes he witnessed tongue-speakers using words that the tongue-speaker had previous knowledge of or education in. He cites Greek and Spanish among these words, phrases or syllables he can recognize in his studies. When this occurs, Samarin believes that the tongue-speaker, in order to convey a certain meaning in the prayer language, such as happiness, or the name Jesus, simply takes words known to him from foreign languages or perhaps his native language and changes them slightly, causing them to sound as if they were a new or different language. Often times, speakers feel like they use more than one language or dialect of a language while speaking in tongues. These perceived deviations and prayer language are recognized by the speaker due to changes in accent pattern, intonation, or the different positioning of the tongue in the mouth during the speaking of the prayer language.

Dr. Samarin analyzed samples of speech for these factors and was able to distinguish several distinctly different patterns of speech from some individual tongue-speakers. He attributes these changes to the production by the tongue-speaker of different sound patterns and does not feel that they are new or varying languages. He also believes that changing emotional states may cause the same sounds to appear differently due to changing intonation and syllable stress patterns during the emotionally charged state of mind of the tongue-speaker.

A very important aspect to be considered in the discussion of glossolalia is whether or not it can ever be proven to be a known earthly language. If the Bible description of speaking in tongues as being "of men and of angels" is correct, then at least some portions of tape-recorded glossolalia should be recognizable as an earthly language.

Dr. Samarin argues, in fact, that many of the instances in which tongues were perceived by listeners as being an earthly language is due merely to superficial similarities between that which was spoken and a foreign language that the listener may have had some rudimentary knowledge or experience with.

He postulates that it may be possible for certain sounds prominent in the language, for example, Russian or French, to be prominent in a portion of tongues and therefore fool the listener who has only limited experience with a foreign language, into believing that he is only hearing that foreign language spoken. Other times, certain sounds in tongues may actually be sounds similar to certain known words in a foreign language. This would therefore convince the listener that he is actually hearing a complete message in the foreign tongue he knows.

Another factor which may cause the tongues to sound like a true foreign language may be the certain inflections of speech used causing the phrases to sound very similar to inflections used in the pronunciation of certain foreign tongues. Samarin has suggested that people with only limited knowledge of certain foreign languages might be easily deceived by something that sounds similar to the foreign language. He and other researchers suggest that the only true methods of ascertaining whether a tongue sample is a true foreign language or dialect of a foreign language is when a native speaker can identify the passage and tongues has a distinct and understandable example of his own native language.

Other important types of voice modulators such as cadence, rhythm and voice tone are important, as is inflection. All of these factors of inflection, voice modulation, cadence, rhythm and voice tone may seemingly imitate that found in earthly languages. For example, many Americans would describe an Oriental language as having a singsong character to it because of the rising and falling intonations.

Therefore, if they were to use that same pattern of intonation in their tongue speaking, it may influence them to think that they are speaking in some Oriental language. A more staccato type of pronunciation of syllables might influence tongue speakers to believe that it also is an Oriental tongue.

Now that we have examined some of the arguments about how tongues may cause listeners to be fooled into thinking that they are speaking a foreign language, we need to more fully explore the question of whether glossolalia has ever been proven to be a real foreign language. This phenomenon is known as zenoglossia.

Zenoglossia is used to describe speaking in known earthly foreign languages. It is a widely held belief in Pentecostalism that while tongues is primarily a foreign language, it has sometimes been used by the deity to tell a message or to serve as a sign to certain persons or perhaps a native of a foreign land in which the tongue's message is given.

Evidence of this belief is found in the Book of Acts when people from different lands heard messages from God in their own language during the speaking in tongues. Other evidence is also found in Pentecostal literature since the turn of the century in which it is reported that foreigners were able to understand messages in tongues during the Azusa Street revival.

One important aspect, which must be examined, is the argument of many, which deride these alleged messages in tongues. Many people find it hard to believe that these messages could be from God and given in an earthly understandable language. This is due to the fact that if this occurred, it would obviously be a miraculous supernatural manifestation. In this discussion, we have seen that there are many ways in which scientists try to explain these seemingly miraculous supernatural phenomena away. However, the truth of the matter may lie in a close examination of sufficient samples of tongue speaking to determine whether or not zenoglossia does occur.

It seems that very little research has ever been accomplished to ascertain if glossolalia has ever been a known earthly language. Scientists seemed to have assumed since some of the earlier studies decades ago were unable to produce evidence of zenoglossia that it just does not exist and is not worthy of their research efforts. Perhaps another method to discover the truth behind the zenoglossia question would be to collect cases in which the native speaker of a language could witness tongue speaking and attest that the utterance was a fluent and understandable example of a known language.

Another phenomenon to explain is how one may speak in a foreign tongue that he doesn't know. This is known as cryptomnesia. Cryptomnesia is a phenomenon in which a person at some time has come into contact with a foreign language such as during childhood while hearing an uncle from another country speak or while hearing another language on television. Perhaps a period of stress can bring forth the memory of this language, which has been held in the subconscious. This hidden memory may be brought from the subconscious mind to the surface and be expressed as a vocalized language.

We can now see how the tongues experience is viewed by one prominent linguistic scientist, Dr. Samarin. He feels that due to social linguistic factors consisting of several modes of motivation such as the motivation performed by a religious command (such as being told that you need to speak in tongues to be in the presence of God) and motivation by the group to become initiated into an activity the group shares in common, that these motivations cause the subject to somehow develop groups of sounds which imitate a foreign language.

Dr. Samarin summarizes his view of samples of glossolalia he analyzed from tongues recordings in the following manner: "They always turn out to be the same thing, strings of syllables made up of sounds taken from among all those that the speaker knows, put together more or less haphazardly, which nevertheless emerge as word-like or sentence-like units because of realistic, language-like rhythm and melody."

He therefore feels that although there are superficial similarities between tongues and real languages, the fundamental truth is that tongues are never a real language. He states that not only has he never been able to ascribe any meaning to tongues but that he has not heard or seen any scientifically documented occurrences of zenoglossia among Pentecostals. Therefore, he feels that instead of being a supernatural act, tongue speaking is a natural occurrence where the human mind subconsciously, and perhaps at times even consciously, produces something that sounds like a language in response to certain motivational factors to do so.

One way to determine if repetition of certain sounds occurs during tongue speaking is to analyze certain transcriptions of glossolalic tests and run a statistical analysis on it. During the statistical analysis each different vowel, consonant and sound is compared to the statistical analysis of the pattern of vowels, consonants, and sounds of the normal speech. This was performed on a text submitted from a charismatic Presbyterian minister.

Dr. Samarin found that there was a dominance in sounds produced by using the tip of the tongue. This represents 57% of all consonants that the speaker used to initiate syllables and it was noted that a similar sound normally represents about the same percentage of sounds in the English language. Dr. Samarin therefore concluded that the Reverend was using a common sound of his native language, in this case, English, to produce tongue speaking.

Other studies have shown that forty to fifty percent of sounds in glossolalic utterance consist of certain sounds that in normal English speech comprise only about twenty-five percent of the total types of sound present.

Another way to analyze a person's speech is referred to as paradigmatic analysis. Paradigmatic analysis refers to understanding what different sounds a speaker knows and is able to choose from and use to produce his speech. In other words, this is a description of the speaker's linguistic resources. Dr. Samarin believes that after analyzing several glossolalic texts, that many lines the tongue-speakers use are really produced from a few basic sounds. The tongue-speaker uses different syllables that are repeated in different patterns to produce the speech.

Dr. Samarin believes that the resources of different sounds, primarily consonants and vowels used by tongue-speakers, are normally those from native language. He therefore believes that people take basic sounds from their own language and use them in different ways such as changing slightly the pronunciation or using them in different patterns to produce something sounding like a foreign language. He also suggested that people may actually be regressing in their language.

In regression, the tongue-speaker, while speaking in tongues, actually reverts back to the most simple and easiest sounds that they have known from childhood. Perhaps those same sounds are those they used as a child when they were babbling their first attempts at using their culture's language.

He also suggests that knowledge of a foreign language either from casual or academic experience may enable the speaker to use various words or sounds from the foreign languages to produce the glossolalic utterances. This, he argues, may be one of the primary ways in which tongue speaking seems to be that of a foreign language.

It is interesting to note that if this idea is true, then when glossolalic prayers are analyzed, there should be striking similarities between the prayer language and either a foreign language that they have already learned, or the tongue-speaker's normal pattern of speech in particular.

One common and easy way to elucidate features of a person's speech pattern is the dialectic idiosyncrasies present in the speaker's native region. For example, a Southerner has certain ways of pronouncing certain vowel sounds and has certain inflections of the voice that can be very dissimilar from a person from another area of the country.

Dr. Samarin examined many recordings of glossolalic speech and was unable to find any examples of consistencies in dialectic patterns in the samples he studied. He was surprised to find that even individual subjects from dialectically distinct regions of the country, for example, Texas, did not pronounce sounds with the same accent while speaking in tongues. Therefore, although their casual speech made it evident that they were from Texas, their tongues or prayer language alone would not have suggested that to be the case.

Language is a means by which man can communicate meaning, and his thoughts and desires to other individuals. It would, therefore, be interesting to try to ascertain what types of meaning might be present in glossolalic speech if one takes Dr. Samarin's thesis that glossolia is a random amalgamation of sounds produced with the intent to sound like a foreign language. If this is so, then there should be no real meaning conveyed by tongues.

Conclusion

The majority of the text in this chapter has been devoted to an historical and scientific discussion of one very intriguing example of how the spirit influences the mind, and then how the mind can profoundly effect our physical abilities and actions. Speaking in tongues and related super-normal acts taken in the physical realm of our being evidence to us the profound impact that mind and spirit can have upon the body.

I believe that disease can also have such an impact. It is interesting to note that many of the physicians with the greatest experience in using "holistic" therapies for cancer have been quoted as saying that improving the mental, spiritual, and psychological health of their patients has been of paramount importance. Surely, each of us must, in our own way, address our spiritual and psychological health if we are to enjoy the full measure of a healthy lifespan.

Chapter Twelve
ANTI-AGING MEDICINALS

This chapter presents medical drugs that have broad scopes of actions as anti-aging medications. It also discusses dementia, and highlights those anti-aging drugs that have been found useful in combating this "mind crippler" that is so common as we age.

Combating aging of the cognitive "thinking" brain and the lower portions of the brain has special importance in anti-aging medicine. The cognitive functions, found in the cerebral cortex, are what give meaning to life and make us each an individual personality. The lower areas of the brain, such as the brain stems, are essential in controlling and maintaining basic physical processes such as motor coordination (muscle control), and sex drive. The preservation of these brain functions and many other essential functions of the body can be significantly influenced by the medications discussed in this section. Let's first consider dementia.

Dementia is a process by which a deterioration of the brain's ability to function normally occurs. This deterioration can occur suddenly or, as with the normal aging process, can occur as a gradual deterioration. Dementia results in a decreased ability to think, to reason and to remember. This deterioration in the cognitive functions of the brain is usually the result of abnormal physical processes that damage the tissues of the brain or interfere with the brain's functions.

One of the primary types of cells in the brain is the "neural" or "nerve cell." These cells, called "neurons," have a large central cell body with extensive elongated projections from this cell body, which interact with other associated neurons. Interconnecting webs of these neurons act together as processing groups just like the central processing unit of a personal computer. These neurons function together to form groups that process thoughts and retain and process memories. The areas where the neuronal cells interact with each other are known as "synapses" and occur at the ends of the elongated processes, which extend out from the main cell body of each neuron. At each synapse, a chemical transfer occurs. This transfer of chemical messengers is the process by which each neuronal cell can interact with the other neuronal cells in its web. These interneuronal chemical messengers are known as neurotransmitters. There are many types of neurotransmitting chemicals. They include norepinephrine, acetylcholine, dopamine, and serotonin.

As neuronal cells interact with one another, the following process occurs. An electrical message is sent down the long processes of the neuronal cell to its end point. This end point, as the synapse, then releases the appropriate neurotransmitter chemical, which crosses the synaptic gap and stimulates the next neuron. An electrical signal is then sent down the next neuron where it is once again processed and sent into the web. This is how the cells of the brain interact together in processing webs that create, manipulate, and store thoughts and memories.

Physical breakdown in any of the components of the neural web can lead to dementia. Therefore, death or injury to the neurons or their long projecting extensions can lead to dementia. Likewise, any diminution in the ability of each neuron to transmit an electrical impulse down its elongated processes may also lead to decreased functioning of the brain.

Dementia can also be caused by disruption in the release of chemical neurotransmitters at each synapse. This may result in the decreased ability of the next neuron across the synapse to receive and process those chemical signals.

Alzheimer's disease is a destructive process by which the brain cells are interfered with to such an extent that they form what are known as "neurofibullary tangles." These tangles are defective webs of brain cells that have become so unhealthy that their ability to form normal connections between themselves is significantly disrupted. As the process continues, the actual neuronal cells themselves die and are eliminated from the body. Therefore, in people with advanced Alzheimer's disease, autopsies of the brain show empty areas where the cells have actually died and been removed.

Adjacent to these empty areas, neurofibullary tangles may be found. Neurofibullary tangles are nonfunctional or partially functional areas of the grossly disrupted neuronal webs. Alzheimer's dementia has been associated with many potential causes. Perhaps the most famous of these is the metal, aluminum.

It has been known for quite some time that neurofibullary tangles contain abnormally high amounts of aluminum. This has led some scientists to speculate that aluminum may be a causative factor in Alzheimer's disease. Although it is certainly, at this point, unknown whether aluminum is a causative factor, it is a well-known fact that many toxic heavy metals can cause brain disease, dysfunction and dementia.

Perhaps the most famous of the heavy metals which cause dementia is mercury. Industrial workers of the last century who were exposed to high levels of mercury commonly developed dementia and were, in many instances, considered "mad." The term "as mad as a hatter" has to do with this process. In the last century, hat-making used large amounts of mercury. Workers were then exposed to it. Unfortunately, the hat makers often became crazy and therefore the term, "mad as a hatter" developed.

In the normal aging process, a slowly progressing dementia is very common. Even in those individuals for whom this process does not ever develop to such a degree that they become hospitalized or placed in a nursing home because of the dementia, it is extremely common that a gradual decline in mental functioning and memory functioning occurs. This gradual decline in the ability to think and remember is common with normal aging and leads to a decreased quality of life even if it does not reach such a drastic stage that a doctor would diagnose it as a "dementia."

Medication can be used to influence both the rate of development of dementia and as a treatment for existing dementia. Medications can act to decrease the deteriorating processes, which slowly lead to dementia. There are many different mechanisms of action of the various anti-dementia drugs. These mechanisms may include both antioxidant effects on the brain and cleansing the brain of abnormal substances such as heavy metals and lipofuschin.

Lipofuschin is a substance that collects in brain cells and other cells of the body during aging. Lipofuschin may consist of accumulated waste byproducts of cell metabolism. Age spots, the darkened areas on the skin, are cells that have accumulated large amounts of lipofuschin. Lipofuschin is a colored pigmented substance, which can be seen on the skin as age spots. When lipofuschin accumulates within the brain cells, it may be a causative factor for dementia. It may also be a factor in the gradual deterioration of the brain functioning with age.

Medications can also work systemically throughout the body to decrease the aging process. Aging of the cells of the brain can lead to progressive dementia. Aging of the other cells of the body can lead to deterioration of the physical stamina and functioning of the body as a whole, as well as specific deterioration of different organs and organ systems of the body.

It is well known that as we age, our ability to cleanse the blood through kidney filtration lessens. Medications, which act systemically on a wide variety of organs including the brain, may help to decrease deterioration of other organs such as the kidneys, the lungs, the heart, and the hormone-producing glands.

Deprenyl

Deprenyl also known as "Eldepryl" is the brand name of a powerful anti-aging drug also known as Selegeline. Deprenyl is approved for use in the United States for the purpose of treating Parkinson's disease.

Parkinson's disease may be very similar to the basic normal aging process that occurs in the brain. In Parkinson's disease, cells deep in the brain, which control movement and fine coordination, are slowly destroyed. This results in severe muscular instability, tremors, and other difficulties involving muscle control. These same cells, located deep in a part of the brain known as the substantia nigra, naturally deteriorate slowly with age. The slow deterioration of these cells is thought by anti-aging scientists to be a primary cause of the Parkinson-like symptoms that occur very commonly in the elderly.

It is well known that if people are fortunate enough to reach advanced age, it is very common for them to lose much of their muscle coordination and to have fine tremors. It has also been noted by anti-aging scientists that those people who have better survival of the functioning of the substantia nigra cells also tend to live longer.

The cells of the substantia nigra are dopamine-producing cells. Dopamine is the neurotransmitter that these neuronal cells in the substantia nigra used to connect with each other via their synapses. It is thought that as dopamine breaks down or is metabolized, it forms 6-dydroxy dopamine, which may generate oxidative free radical damage.

It is postulated that the free radical damage caused as a natural result of normal dopamine metabolism slowly poisons the cells of the substantia nigra, which are producing the dopamine. In this instance we find an example where an essential neurotransmitter, dopamine, may actually break down into a substance which slowly poisons the very neuron cells that it aids as a neurotransmitter.

Scientists believe that this process is slowly occurring throughout our life. By a certain age, humans will have developed such a significant loss of these cells in the brain that they will start noticing symptoms of deteriorating muscle coordination and the development of tremors. It is estimated that roughly eight percent of the neurons of the substantia nigra must be destroyed before significant symptoms are noted. However, many anti-aging scientists believe that these same cells, deep in the brain, play roles in many other areas of human health aside from muscle coordination and activity.

Deterioration of the substantia nigra and related neurons is known to cause symptoms such as decreased sexual functioning and general physical debility. In scientific studies in which various strains of laboratory rats have been given Deprenyl injections, it was found that the rats lived to much greater lifespans than did the rats not given Deprenyl. In addition, the rats given deprenyl appeared healthier, with better body weight and shinier coats. As a matter of fact, at this time, it appears that Deprenyl produces the greatest addition to lifespan of any known and tested medical therapy used for life extension purposes. The results of one of the major studies with rats, if extrapolated to humans, would predict a human lifespan of 150 years. This is a dramatic increase in the normal human lifespan, which is generally 100 years or less.

Deprenyl can have unpleasant side effects, especially when taken in higher doses. High doses are commonly prescribed for people with Parkinson's disease. However, as a life extension medicine, smaller doses are probably more effective and wiser. Chronic low dose therapy of Deprenyl starting at age forty-five or later may help to decrease the aging of the brain. It is common practice among life extensionists to take five mg tablets of Deprenyl every other day. Likewise, much smaller daily doses have also been used. These daily doses range from 1.8 to 2.5 mg daily.

This low dosage range is usually very well tolerated. Undesirable side effects such as nausea are seen in ten percent of test subjects. Dizziness, light headedness or feelings of being faint are seen in seven percent of test subjects. Abdominal pain, confusion and hallucinations are seen in less than four percent of subjects studied. It is anticipated that the lower dosage regimens used for life extension will result in a far lesser percentage of adverse reactions than those just quoted.

It should be noted that deprenyl should not be used at the same time a person is taking opiate drugs such as Demerol. Deprenyl is available as a prescription drug named Elderpryl and is marketed by Sommerset Pharmaceuticals in 5 mg and 10 mg tablets. The 10 mg per day dosing is a regimen that is far more likely to produce the adverse side effects that were previously mentioned. This dose of more than 5 mg per day seems both unnecessary and unwise for those seeking to use Deprenyl for its life extension activities.

Hydergine

Hydergine is a very potent anti-aging and anti-dementia drug. Hydergine, a brand of ergoloid mesylate, is marketed as a prescription medicine in the United States as oral tablets, sublingual tablets and in liquid form. The oral and sublingual, as well as the liquid capsules, are available as one mg of Hydergine per tablet or capsule.

Hydergine has been found to be very helpful in individuals over sixty years of age who manifest indications of decreased mental capabilities including difficulty with thinking, poor mood, difficulty with caring for themselves, lack of motivation, and decreased ability to deal with others. In young individuals or individuals of any age interested in anti-aging therapies, Hydergine is a very powerful tool in a life extension program.

Hydergine seems to act in many ways. Some of the actions of Hydergine are to, 1) increase the brain's blood supply, 2) improve the oxygen delivered to the brain, 3) enhance the brain cells' ability to metabolize oxygen and other substances, and 4) protect the brain cells from damage due to decreased oxygen supply such as found in strokes or in people who have deficient blood vessels within the brain.

It also has been found to decrease the deposition of lipofuschin, the aging pigment, within brain cells. It helps to prevent damage to the brain cells by oxidative free radical damage and it has been found to improve learning abilities, memory function and general intelligence. Hydergine is therefore one of the most significant and powerful "nootropic" agents or "smart pills."

Nootropic agents are those medicines that are known to improve the cognitive functioning of the brain. These drugs help us to seem smarter. In other words, they help us to think more clearly and to improve our memory functions and our ability to recall facts that have been learned in the past.

Hydergine has been found to be extremely well tolerated with an extremely low incidence of any type of adverse reactions. Hydergine can, however, cause bradycardia (slowed heart rate), rashes, nausea, drowsiness, and when used with caffeine, can cause headaches or insomnia). The main contraindication to using the drug would be in those who are known to be allergic to it or in people who are suffering from psychosis.

Practitioners of life extension use Hydergine in different dosage ranges. It is commonly used in dosages ranging from one mg per day to 10 mg per day. The common dosage of Hydergine recommended by the manufacturer for the treatment of senile dementia is one mg three times per day. A total daily dosage of six milligrams per day has been shown to induce growth hormone secretion in the elderly, who are the least likely of all age groups to respond to growth hormone releasing substances.

Deaner

The prescription drug Deaner is no longer available in the United States. It is marketed in Europe under the name Deanol-Riker. Deaner is a well tolerated medication that can have marked effects to improve brain functioning. It improves learning and memory skills and aids in concentration and attention span.

These actions are thought to result from the activity of "DMAE" found in the drug. DMAE can also be obtained as "DMAE bitartrate." Commonly prescribed dosages range from 200 mg per day to more than 1000 mg per day, with a common effective dose being 400 mg per day. Side effects may include insomnia, weight loss, and headaches. Persons suffering from any kind of convulsive disorder or "epilepsy" should not use Deaner.

GH3

Gerovital GH3 is a medicine that can act systemically to decrease aging of various cells and organs of the body. Gerovital is a special form of the commonly used medicine Procaine. Dr. Anna Aslan from Romania developed Gerovital from Procaine and started using it in 1951 to treat elderly patients for the debilities of aging. Gerovital is a two percent solution of Procaine Hydrochloride and contains small amounts of potassium, metabisulphate, disodium phosphate and benzoic acid. Experience in using the medication by Dr. Aslan and other physicians over the last several decades has lead them to conclude that it is widely beneficial for many of the processes which occurs with aging.

It has been found to have wide-ranging effects on the body. It benefits aging dysfunctions such as graying of the hair, wrinkling of the skin, sexual dysfunctions, poor memory, lack of energy, depression, arthritis, heart problems and tremors. Laboratory animals were found to have a markedly prolonged lifespan when given Gerovital GH3. Dr. Aslan has studies showing an increased life span in male white rats of twenty one percent and for females of six percent.

As Gerovital is processed by the body, it is broken down into at least two different metabolites. These are DEAE or Diethyl amino ethynol and PABA, para amino benzoic acid. PABA is a known member of the B vitamin family and DEAE is related to a known membrane stabilizing medicine. DEAE has been found beneficial in the treatment of decreased mental functioning of the elderly. DEAE used for this purpose is known as chlorpromazine. DEAE, one of the metabolites of Gerovital, is very similar in structure to DMAE, which is the primary element of the anti-aging and anti-dementia drug known as Deaner.

Due to the fact that Gerovita is a form of Procaine, people should be cautious if they have been allergic to anesthetics related to procaine such as novocaine, lidocaine and procaine itself. Gerovital is available in several different forms including tablet, liquid or injections.

Piracetam

The noted nootropic agent Piracetam (Nootropil) can aid in many cognitive functions of the brain. Not available in the United States, this medicine is popular in other countries and is generally well tolerated. It is reported to enhance intelligence, memory, the ability to concentrate, and creativity. Nootropic drugs related to piracetam include pramiracetam, oxiracetam, tenilsetam and aniracetam.

Piracetam and its relatives are thought by some researchers to be directly beneficial in the treatment of a variety of dementias and Alzheimer's disease. It is available in 400 and 800 mg units and it is commonly taken three times a day. Lesser doses may be effective when given with other nootropic agents such as Hydergine. Side effects may include nausea, sleep disturbances, headaches, and gastrointestinal distress.

Thymosin

Thymosin is a hormone substance meant to replace the family of hormones produced by the thymus gland. The thymus is a gland located under the breastbone near the base of the neck. It is large during childhood when it is very active in helping to eradicate infections and in building up the immune system.

Around the time of puberty, the thymus gland begins to shrink or "involute." This involution of the thymus gland proceeds over the adult lifespan and is thought to be a major factor in the gradual deterioration of the immune system as we age. As the thymus becomes smaller and involutes, it produces less of its hormones known as "thymosin hormones."

Since the 1960s, clinical studies of thymosin hormones have been undertaken. It has been found to be useful in improving immune system functions in humans and has been used as a cancer treatment. Replacing thymosin to levels more like they were in young adulthood is thought to help reinvigorate the immune system. Thymosin may prove useful as a general acting anti-aging drug as well as having a beneficial effect on the immune system.

Vasopressin

Vasopressin or "ADH" is a potent anti-aging drug marketed in the United States as the prescription drug "Diapid." Vasopressin is a natural hormonal substance produced within our brain. It is made in the anterior area of the hypothalamus portion of the brain and then is released into the body from the posterior portion of the pituitary gland.

Vasopressin has many influences but is particularly noted for its effects on kidney function, the metabolism of carbohydrate, and its influence on learning and memory. This hormone is called ADH or antidiuretic hormone due to its effects on the kidneys. It helps the kidneys to conserve body water. Animal experiments have shown that vasopressin can keep laboratory mice from developing amnesia when exposed to electrical shock. It can also improve or even reverse memory dysfunctions in laboratory animals and can improve the learning ability of test animals.

Vasopressin seems to effect the neuron cells of the brain in ways that improve their ability to form patterns that create memories. In so doing, it appears to improve the brain cells' ability to form long term memory. The hormone also seems to be beneficial in assisting the brain to recall stored memories.

In addition to the animal studies, experiments with human patients have also shown the beneficial effects of Vasopressin. Patients suffering from injury-induced amnesia were found to have significant recovery of memory functions after Vasopressin treatments. Experiments with normally aging people have shown that their memory functions can be significantly restored with Vasopressin. Treated patients had better reaction times, better memory, greater attention spans and better concentration.

The results of the many studies done with Vasopressin seem to indicate that it would be more effective in people who do not yet have significant brain disease and mental dysfunction. Vasopressin is available as the prescription drug Diapid in nasal spray form. One spray delivers approximately 2 USP units of the drug to the nose. Normal dosages are one or two sprays of Diapid, one to four times daily. Life Extension Biotherapies recommends the use of Diapid as one or two sprays in the late afternoon and then again at bedtime. Diapid should be used with caution, if at all, in patients who have significant cardiovascular disease. Diapid may act to increase the blood pressure and cause other symptoms that could be problematic in people suffering from disease of the arteries and heart.

GHB

Gamma hydroxybutyrate is a natural chemical that is related to the neurotransmitter of GABA. Currently under investigation by researchers at the University of Chicago, this powerful agent appears to have significant effects on the neuroendocrine system. It increases growth hormone release, induces sound sleep, and reportedly also improves sexual functioning. At least one study has found that it also increased prolactin levels. Prolactin generally is inhibitory to the secretion of sex hormones. Any increased production of prolactin may therefore interfere with many of the anti-aging therapies discussed in this program.

GHB is no longer available in the United States. After being abused by body builders, it was removed from the domestic market by the FDA. Life Extentionists have used GHB in doses ranging from 500 to 1000 mg at bedtime. Potential side effects, particularly when taken in high doses, include muscle spasms, muscle fasciculations and an overly deep state of sleep. Research studies currently underway are expected to give us much needed information about its actions, side effects, and potential drug interactions and contraindications.

Homeopathic Remedies

Homeopathic remedies work differently than regular medicines. Conventional medicines work by causing a chemical reaction between the medicine itself and some chemical component of the human body. Homeopathic medicines work by causing the body's own defensive energy pattern, called the "vital force," to become stronger and to therefore cause the body to ward off the illness or disease.

Homeopathic remedies commonly use minute quantities of some herbal or other substance which have been specifically processed to stimulate the body's vital force so that it can rid itself of the disease process. There are quite a few homeopathic remedies that have been found useful over the years in combating the aging process and medical problems directly associated with the basic aging process.

Baryta Carbonate

Baryta carb or carbonate of baryta is a powerful homeopathic remedy for elderly people, especially elderly men. It is particularly effective in cases of senile dementia, cases of heart and blood vessel disease involving fibrosis of the arteries, deterioration of the arteries with aneurysms and ruptures of the arteries. Baryta carb, therefore, is most indicated for the senile deterioration of elderly men, particularly when the degenerative changes are just beginning and involve cardiovascular and cerebral or brain problems. The homeopathic remedy baryta carb is usually slow acting and takes a repetition of dose until its effect takes hold.

Orchitimun and Oophorinum

For members of both sexes suffering from sexual weakness secondary to generalized senile decay, homeopathic extracts of the ovaries and testes may be beneficial. For men, Orchitinum, a homeopathically prepared testicular extract can be used. For women, Oophorinum may be helpful. Oophorinum is also beneficial for dysfunctions due to menopause and to other generalized climacteric disturbances. The presence of a skin disorder, acne rosacea, may also help point to the potential benefits of this remedy for sexual weaknesses secondary to senile decay.

Combination Remedies

Combination remedies are remedies that combine several homeopathic remedies into one single treatment. These combination products have become much more popular in recent years among those who use homeopathic medications. Combination remedies utilize several homeopathic medications which together seem to have a greater beneficial effect on certain disease states and symptomatoligies than do individual homeopathic remedies. Combination remedies also make it easier for the non-physician to self prescribe. It is more difficult for a non-physician to ascertain the exact single homeopathic remedy that would be most beneficial for their condition than it is to choose a beneficial combination remedy.

Personal Constitutional Remedies

Constitutional remedies are those homeopathic treatments that seek to treat, at the deepest level, the unhealthy or weak traits of personality and detrimental physical attributes. Constitutional remedies work at the most basic level of the vital energy of the body, inducing the vital force to throw off unhealthy processes both of the mind and of the body. Constitutional remedies are an individualized treatment. It takes a significant amount of time by a qualified homeopathic physician to deduce the correct prescription.

Although the personal constitutional remedy may not have a direct or immediate effect as an anti-aging treatment in and of itself, the basic fact that the constitutional remedy will "strengthen the organism as a whole" will be of benefit to any anti-aging program. It will benefit the anti-aging program because if the organism, as a whole, is at its healthiest and most optimal state, then treatments that are specifically aimed at reducing or reversing the normal aging process will have a better chance of success. Individuals whose constitution is weak and sickly will tend to age more rapidly than they would have otherwise.

Homeopathy generally considers the aging process to be a natural and normal process, which the human body is designed to undergo. Lifextension Biotherapies believes that this normal process of gradual deterioration that we know of as "aging" can be modified and potentially even reversed through proper anti-aging medical treatments.

Therefore, even though it may appear at first glance that homeopathic constitutional therapy is at odds with the idea of anti-aging treatment, this may not, in fact, be true. Homeopathic constitutional therapy, in and of itself, will strengthen the body and mind to the highest level that it was designed for. If, in addition to this, anti-aging treatments are added, an even more dramatic benefit can be expected and the aging process can be modified.

SECTION FIVE - ALTERNATIVE CANCER TREATMENT

Chapter Thirteen
ALTERNATIVE CANCER THERAPIES

Breakthoughs in Natural Cancer Care

During the past half-century, an explosion in new and innovative alternative, natural cancer therapies has occurred. Although these natural concepts are most often completely rejected by the conventional bio-medical establishment, they have survived and flourished. The phenomenon has now reached the point that conventional medical organizations are slowly beginning to embrace the same alternative therapies they formerly so thoroughly despised.

Early alternative cancer modalities revolved around stimulating the immune system, detoxifying the body, stimulating and rectifying the "vital force," and the use of simple natural substances such as high-dose vitamin C. More recent developments utilize the healing power of the body in newer, more profound and more varied ways. The most promising of these "breakthrough" natural cancer treatments includes R-A Therapy and the use of telomerase treatments to fight malignancy. This chapter will explore many intriguing alternative cancer therapies, some of them old standards, and others the newest and most advanced the field of natural cancer care has to offer.

R-A Therapy

Perhaps the most important breakthrough in cancer medicine, both alternative and conventional, is the concept of inducing the cancer cells to "heal" themselves. For many decades, conventional chemotherapy, surgery and radiation therapy has sought to "remove" the tumor via surgery, or to "destroy" tumors via radiation, or to "kill" the cancer via chemotherapy.

Alternative medicine has similarly used external forces (those outside the cancer cell itself) in attempts to effectuate healing. Alternative techniques basically try to help the body heal itself of cancer, rather than trying to induce the cancer cells to heal themselves. Alternative therapies such as immune system stimulation and detoxification seek to strengthen the natural defense mechanisms of the body as a whole, in an attempt to get the body as a whole to rid itself of the cancer. Now, however, both alternative and conventional cancer medicine are discovering the dramatic results possible when they focus attention on ways to cause the cancer cells to attempt to "heal" themselves.

The "R" in R-A Therapy stands for Re-differentiation, the process by which cancer cells may be induced to revert to a non-cancerous state. Recent biomedical discoveries have shown two new mechanisms by which cancer cells may be able to change their very character, and in essence "heal" themselves of being malignant. One of these mechanisms is called re-differentiation.

Re-differentiation is an internal process within the cancer cell. It involves causing the genes of the cancer cell to instruct the cancer cell to revert to a non-malignant condition. Genes contain the instructions for cell form and function. Genes direct the activities of the cell and control every aspect of the cells' growth, development, activities, and lifespan. Laboratory research has demonstrated that cancer cells may be able to sense the presence of external stimuli (such as certain natural chemicals) and respond to the stimuli by changing the shape, character and activities of the cancer cell, causing the cancer cell to revert to a non-cancerous state.

The reason this can occur is that cancer cells are descendants of normal, healthy cells. When normal, non-malignant cells are exposed to stressors, they may slowly mutate until they have become so profoundly abnormal that we categorize them as "cancer." Scientists believe this transformation from normal to malignant occurs in many steps. As each step toward malignancy is taken, the genes of the cell mutate, and they then instruct the rest of the cell to take on the abnormal form and function of a mutated cell. When a certain level of genetic mutations has taken place, the cell is a full-fledged cancer cell.

An example of the step-wise transformation of normal cells into malignant cells is found in examining PAP smear reports. A PAP smear is a small sample of cells from the female cervix. The smears of cells are viewed under a microscope to determine their characteristics. The cells can appear like normal, healthy cervical cells, or they may be seen to be abnormal. Abnormalities can range from mildly abnormal, moderately abnormal, and finally culminating in cells so severely abnormal that they are officially called "cervical cancer."

When women have mildly or moderately abnormal cells, they are watched closely to ensure that the abnormal cells do not continue transforming themselves into cancer cells. As time goes on, the abnormal cells may continue mutating until they are cancer cells. As each step in the transformation of normal cervical cells into cervical cancer cells occurs, the genes of the cells make step-by-step progress towards becoming cancer. It is thought that these steps are "remembered" by the genes of the cell, so that cancer cells may have a "genetic memory" of all the steps they took from a normal cell to a malignant cell.

Re-differentiation occurs when the cancer cells are induced to take some, or maybe even all, of the steps backward. In other words, re-differentiation occurs when cancer cells take steps backwards toward the way they were in the past, toward becoming, once again, a normal non-cancerous cell. Cancer cells may take all the steps back and revert into a non-malignant state, or they may take some of the steps back and become less malignant, although not entirely normal. In either event, reversion of the cell towards normalcy makes the cells less cancerous, and less dangerous.

The "A" in R-A Therapy stands for apoptosis, the process by which cancer cells may be induced to "kill" themselves. Apoptosis is the scientific word for "cell suicide." Cells may kill themselves, commiting suicide, when they detect that they are grossly abnormal. They do this by using a genetic system that senses the cells have become abnormal, and then proceed through a pre-programmed process of self-destruction.

Certainly a cancerous cell is grossly abnormal, so why doesn't the system of apoptosis eliminate cancerous cells? The reason is simple. For cancer cells to develop, the program for apoptosis must be turned off or in some other way disabled. Mutations that disable the apoptotic program are essential components in the transformation of normal cells into cancer cells.

You have probably heard that cancer cells are constantly forming in our body, but are eliminated by the immune system. I believe that the immune system does not eliminate them. I think that the apoptosis sytem eliminates them. In other words, as cancer cells are developing, they eliminate themselves via apoptosis, and it is only the cells that have successfully mutated in a way that turns off the protective apoptotic program that can remain alive to grow into tumors.

If disabling the apoptotic program is essential to development of cancer cells, then we can imagine that turning the apoptotic program back on in cancer cells may cause the cancer cell to "eliminate" itself via "cell suicide." Research is showing that this can indeed occur. As a matter of fact, it now appears that apoptosis is the single most common way in which cancer cells can be killed. Strangely enough, even though we say we are killing them, they are actually "killed" when we induce them to "kill themselves."

R-A Therapy utilizes many natural chemical substances, most of them derived from plants, to induce the cells to either kill themselves via apoptosis, or to revert to a non-malignant, or perhaps a less-malignant state. It is interesting that many of these natural substances also seem to have other substantial benefits to the body. Some act to detoxify, others to stimulate the immune system, and still others may act in beneficial ways not yet understood.

This is why R-A Therapy generally produces beneficial side-effects, instead of the harmful negative side-effects commonly seen with conventional treatments such as chemotherapy and radiation. Unless a person is allergic to one of the R-A Therapy substances, in my experience it is extremely unlikely that they would ever experience any medically significant side-effect from R-A Therapy. These facts make R-A Therapy, in my mind, the single most effective and least toxic system of natural healing in cancer.

People often ask, "If R-A Therapy is such a breakthrough, why don't we hear more about it?" I believe that you will. Alternative medicine is still stuck on talking about detoxification and immune system stimulation for cancer, but now we are beginning to see apoptosis and re-differentiation mentioned from time to time. As the importance of these two processes becomes more obvious, you will hear them mentioned more and more in alternative medicine.

The truly amazing thing, though is how this concept is catching on in conventional bio-medicine. Biotech companies, the major pharmaceutical companies and the major universities are all devoting great time and financial resources to finding ways to utilize apoptosis and re-differentiation.

New drugs are ready to be released to the market that utilize these natural concepts. "Aptosin" is an exciting example of this phenomenon. An older drug used to treat arthritis, Sulindac (brand name Clinoril), was discovered to have the ability to induce apoptosis.

Realizing that fact, a pharmaceutical company patented the portion of the arthritis drug molecule that causes the apoptosis and named it "Aptosin." Their process eliminates the part of the drug molecule that treats the arthritis and retains only the portion that causes apoptosis.

The drug is currently in FDA trials for safety and effectiveness. We can only hope that new medicines, which use this natural means to fight cancer, can be successfully developed. I predict that such drugs will be so nearly natural that they will have few negative side-effects and may gradually replace many of the chemotherapy and radiation treatments currently used in conventional cancer medicine.

Telomerase Treatments

The use of telomerase treatments to fight cancer may also revolutionize cancer medicine. Telomerase is an enzyme found internally within the nucleus (the portion of each cell that contains the genes which control the cell) of each cancer cell. It is not a food enzyme or a digestive enzyme. The fact that it is present and active within cancer cells confers upon them something cellular biologists call "immortality."

Telomerase is said to immortalize cells because it allows cells to reproduce indefinitely. The ability of cancer cells to reproduce copies of themselves for a seemingly limitless period of time makes them much more difficult to kill or eliminate from the body. This is true because they do not "age" or weaken with the aging process, as do the normal, non-malignant cells of our body. The normal, healthy, non-cancerous cells of our body are almost all telomerase-free. In other words, they do not contain telomerase within their cell nucleus and therefore will "age" over time, become weaker and less functional as they age, and will have a limited number of times they can reproduce themselves. This dramatic difference in cellular aging gives cancer cells a tremendous advantage over the rest of the cells of our body.

Conventional Pharmaceutical Telomerase Treatments

Medical treatments to turn-off telomerase activity in cancer cells may make them much easier to eliminate from the body. There are two major ways to attempt to use telomerase techniques in cancer medicine. They are, 1) the use of homeopathic telomerase techniques, and 2) the use of conventional medical and pharmaceutical techniques.

The major bio-technology and pharmaceutical companies are investing heavily in telomerase techniques. They are seeking ways in which they can remove the activity of telomerase from cancer cells. Methods of deleting telomerase from cancer cells may directly, physically, cause the cancer cells to lose the protective and strengthening effects that telomerase imparts to the cancer cells.

Homeopathic Telomerase Treatments

Homeopathy is premised upon the concept that "like treats like." It uses specially prepared versions of natural and chemical substances to improve the functioning of the vital force. The vital force is the homeopathic term for the life force. It connotes the same energy that the chinese call "chi." Rectifying the vital force is thought to ameliorate disease and favor good health.

Homeopathic telomerase is intended to modify the vital force in ways that ameliorate part of the harmful health processes engendered by telomerase, when it is present within cancer cells. We might think of oncologic homeopathic telomerase as an "anti-telomerase" since it is prepared from the telomerase found within cancer cells, and is intended to rectify the derangements in the vital force caused by the telomerase which is present within cancer cells.

The primary difference between conventional pharmacologic telomerase treatments and homeopathic telomerase treatments is that the former works by directly, physically, interacting with the cancer cells to nullify the effects of telomerase within the cancer cells. The homeopathic telomerase treatments are intended to nullify the aberrancies in the vital force, rather than to directly, physically, interact with the cancer cells.

Dendritic Cell Therapy

Dendritic cells are immune system cells that have been strengthened and enlivened so that they may tend to act much more forcefully to work against malignancy. We are all aware that it has been hoped for a long time that the immune sytem can be employed effectively in the fight against cancer. Both in alternative and in conventional medicine, this has been a great priority. Interleukins, interferons, other cytokines, and other immune cells, chemicals and processes are currently in use or under study. Dendritic cells are a further attempt to utilize the powerful healing properties of the immune system to affect cancer. There are at present two major types of dendritic cell therapy. The older version utilizes living immune system cells from the blood of the patient or from donors to prepare the dendritic cell therapy. A newer version uses special vaccines and other modalities that appear to be more effective and convenient than the older method.

Homeopathic Cancer Vaccines

Cancer vaccines, prepared homeopathically, are intended to effectuate the vital force to resist maligancy. Such vaccines are normally administered as injections, into either the layer just under the skin or into the muscle. As with all vaccines, homeopathic cancer vaccine is intended to stimulate the body to use its natural defenses to resist the disease. In this sense, we need to consider the differences between prophylactic vaccines and vaccines utilized during an active disease process.

Childhood immunizations are examples of the former type. They are administered during childhood prior to the time the child is infected with the virus or other infectious agent, in an attempt to foster a very strong immune response against the infectious agent once it is eventually encountered by the child. Vaccine used during active disease are somewhat different in that they attempt to stimulate the defensive mechanisms of the body to resist a disease after the infection or other disease process has already begun. In cancer vaccine therapy, the purpose is to stimulate the resistant forces of the body to fend off the active disease process.

Anti-Angioneogenesis Agents

Anti-angioneogenesis is certainly one of the most convoluted biomedical terms one is ever likely to encounter. But the meaning of this long word is actually quite simple. Anti-angioneogenesis agents are medical agents employed to stop the growth of new blood vessels that feed cancer cells and tumors. All cells within our body require oxygen and nutrients to live and grow. Cancer is no exception. Tumors must have an adequate blood supply or they will die. One way they are able to continue growth is that they send out chemical messengers that stimulate the growth of new blood vessels to the tumor. Factors that interfere with that process are the anti-angioneogenesis agents.

The most famous of these agents found in alternative medicine is shark cartilage. Shark cartilage is used in attempts to stunt the growth of new blood vessels to supply tumors with nutrients and oxygen. Scientists believe that squalene may be the active natural chemical substance found within the shark cartilage that may inhibit blood vessel development.

There are many forms of shark cartilage and squalene supplements available. Perhaps one of the most powerful form is flash frozen live cell shark cartilage. This is normally administered orally after being very briefly defrosted. Liquid flash frozen live cell shark cartilage may represent the freshest and most concentrated form of this widely utilized supplement.

A plant that grows as a wild vine in the midwestern United States shows promise as an anti-angioneogenesis agent. Convolvulus supplements are promoted as natural herbal alternatives to animal-derived shark cartilage. One thing is for sure, if convolvulus becomes the most popular form of anti-angioneogenesis supplement, sharks will be able to rest much easier!

Natural Poly-Phenols

The naturally occurring poly-phenols are a class of natural chemicals that have been renowned for their many beneficial health effects, including activity against cancer, for at least 150 years. One of the leading homeopathic medications for cancer in the nineteenth century was "carbolicum acidum," or homeopathic phenolic acid. Modern research tends to bear out the value of phenols against cancer.

The University of California at Davis is currently conducting research into the many health effects of natural phenols, particularly those found in apples. The old saying, "An apple a day keeps the doctor away" may be, in part, due to the beneficial health effects of the natural poly-phenols found within them.

Another interesting example of the anti-cancer effects of natural poly-phenols is the work of Dr. Stanislas Bryzynski, MD, PhD. His "anti-neoplastons" seem to primarily be various incarnations of phenol attached to other natural chemicals. FDA sponsored research studies at Dr. Bryzynski's medical clinic seem to be pointing to the relative safety and effectiveness of the anti-neoplastons, including phenylacetate and phenylbutryrate.

Phenylacetate and phenylbutyrate are at times administered by injection or intravenously. A powdered form of phenylbutyrate has been administered orally as a supplement. Other forms of poly-phenols and associated natural homeopathic substances used in alternative cancer medicine include homeopathic carbolicum acidum, butyric acidum, acetic acidum, phenylic acidum, catechins, and carcinosin.

Diet and Nutrition

Beneficial food nutrients that may rationally form part of a nutritional cancer biotherapy program include algae, sea weed, certain mushrooms, fish oil, linseed oil, garlic, and fiber. Let's begin with the fatty acids found in fish and flax seed oils. Fish oil and the plant-derived oil known in some circles as linseed oil (flaxseed oil) have been reported as beneficial. Both fish oils and flaxseed oil contain relatively high amounts of an oil, which we don't often consume in much abundance. That oil is omega 3. Omega 3 oils themselves, or a more healthy balance of omega 3 to saturated and omegas 6 oil, may result in changes in cancer cells. Some scientists believe that omega 3 oils are one of the natural substances that have the ability to "de-transform" or inactivate cancer cells. They speculate that this occurs via beneficial changes in the oily "lipid" cancer cells' outer membrane and inner "nuclear membrane."

Many vitamins and minerals, particularly those with strong anti-oxidant properties, are renowned for their use in alternative cancer therapy. Perhaps chief among these is ascorbic acid, better known as vitamin C. This powerful antioxidant can enter all watery environments within our organs, tissues, and cells. The vitamins can act there to quench free radicals so that they are less available to mutate and damage the critical components of our cells. Doses of ascorbate, which are much higher than the government's established RDA (recommended daily allowance), are required to produce significant antioxidant effects. Therapeutic doses vary for each individual, but are generally in the range of several grams per day orally, and much more during IV therapy.

Vitamin A, another powerful antioxidant, is also in the forefront of holistic cancer medicine. It acts primarily in the fatty areas of the cell, such as the outer cell membrane and the membrane surrounding the cells' nucleus. High doses of A, and its sister fat soluble antioxidant Vitamin E, may act to quench free radical damage in the crucial fatty or "lipid" areas of our cells.

Germanium is a trace element, which is helpful to strengthen our resistance to many forms of cancer. This rare element is also being studied by conventional science to fully ascertain its therapeutic role in the treatment of malignancies. It is also important to maintain adequate amounts of selenium and other minerals in our systems as these work together with the vitamin antioxidants to decrease our inherent free-radical activity. Zinc is used commonly in the treatment of prostate cancer, and as a therapy for non-cancerous prostate enlargement.

Fiber, a seemingly simple component of many foods, especially whole grains and vegetables, can have a dramatically beneficial influence on cancer development and progression. Soluble dietary fiber is known to bind excess fats in the digestive tract and cause them to be eliminated from the body in the feces. If they were not removed via the feces, they would be absorbed and enter the blood stream.

Excess blood fats, or fats lying dormant in the colon, may be causative agents in cancer development. Dietary fiber may also act to lower circulating levels of certain harmful hormones. It is conjectured that excess estrogen, as compared to progesterone, may be carcinogenic in some women genetically predisposed to breast cancer. Imbalances of other hormones, in both sexes, may also play a role in cancer. Estriol administration shows promise in cancer.

Food group balancing also plays an important role in a rational alternative therapy regimen for cancer. Depending upon the cancer type and general condition of the patient, variations in the relative amounts of fats, proteins and carbohydrates may prove beneficial. For cancer patients, complex carbohydrates and certain fats (non-animal fats, such as omega 3 and omega 6 oils) may need to be increased in relationship to proteins. Additionally, proteins from animal sources may need to be minimized to allow for more protein from non-animal sources to be consumed. Balancing of food groups should be considered only upon consultation with a qualified physician or dietician familiar with alternative theories of diet in disease management or prevention.

Enzymatic Therapies

Certain enzymes have long been advocated by practitioners of alternative medicine as potent therapies for cancer. Enzymes may work in many ways to strengthen the body's resistance to cancer and to weaken the cancer cells. This occurs via a process that changes the internal milieu of the body, or in other words, the internal body chemistry, to make it a less favorable place for cancer cells to proliferate and to make the body a better place for healthy cells to grow and dominate.

The Laetrile Controversy

Certainly one of the most controversial agents commonly used by alternative practitioners to treat cancer, laetrile or "amigdalin," sometimes also called vitamin B17, is very popular for its reported beneficial effects. Interviews with alternative cancer doctors reveal their positive experience with the substance to include an increased sense of well being, lessened pain and other discomfort, and improved survival in cancer patients. Few alternative cancer doctors refer to it as a "cure" for cancer, but nearly all seem to regard it highly in their regimen of therapeutic options.

Even though most people refer to laetrile, B17, and amigdalin synonymously, there is actually a difference between amigdalin and laetrile. Amigdalin is the name commonly given to the natural product derived from the pits of apricots, while laetrile is a slightly chemically altered man-made chemical version of amigdalin. Scientific research leads us to conclude that intravenously administered amigdalin is safer than that taken orally. This is due to the fact that amigdalin contains natural cyanide molecules, which seem to be released when they interact with the digestive tract, but the naturally occurring cyanide does not seem to be released into the normal healthy tissues when it is administered intravenously. Some speculate that it is the cyanide-containing natural molecules found within amigdalin that is the primary beneficial agent in cancer.

IV Nutritional Biotherapies

Why give nutritional substances intravenously (IV)? It is to dramatically increase the amount of the nutrient that is "bioavailable" to the tissues and organs of the body that need them. Bioavailability is a term describing how much of a substance is actually made available, in a useable form, at the cellular level where it is needed. Nutritional substances, when eaten or taken as oral supplements, must make it past several obstacles before they can be taken up by the individual cells in the body's tissues and organs that need them.

First, adequate digestion of the substance must occur. Stomach acids, gastric and pancreatic enzymes, and interactions with other ingested foods and beverages may all tend to either worsen or improve the digestion of the nutritional substance. Often, the ill patient lacks some components of the digestive process and therefore is unable to adequately digest nutritional substances. Digestion causes the breakdown of nutrients into forms in which they can be absorbed into the blood stream.

The second step is absorption of the digested nutrient into the blood stream. This occurs in different ways for various nutrients. We will not discuss the complicated manner in which some nutrients are transported, but they must all eventually pass through the intestinal wall and into the small intestinal blood vessels. Illness, food and nutrient interactions, and bowel irritability can all hamper the absorption of nutrients into the intestinal blood vessels.

Following absorption, the nutrients are transported through the blood stream to various tissues and organs of the body. Two factors may impair this "assimilation" and transport of nutrients through the blood stream. Many ingested substances flow immediately to the liver before entering the rest of the body. The liver uses its metabolic and detoxification enzymes to alter many ingested substances. In this way, some types of beneficial ingested substances are lost or diluted.

A second factor involves the adequacy or inadequacy of the transport proteins which carry many nutrients and beneficial substances through the blood stream. Defects in these transport proteins may decrease the final amount of usable "bioavailable" nutrients delivered to the cells that need them.

Finally, oral ingestion of many nutrients limits the total amount of the nutrients that can be administered at any given time. Many people, for example, have a tolerance for only a certain amount of vitamin C if taken orally. Beyond this tolerance, their digestive tract may rebel and cause symptoms of nausea, heartburn, gas or diarrhea. Intravenous vitamin C can be safely administered in much larger doses than many are able to tolerate orally.

Intravenous administration of beneficial nutrients can therefore achieve several worthwhile goals. It can allow for easier administration of higher doses, bypass poor digestive processes, bypass inefficient absorption processes, bypass initial liver metabolism of the substance, and present such a large amount of nutrient that poor blood transporter proteins do not significantly affect the amount of nutrient that makes it to its final destination, the cells. By these unique mechanisms, "bioavailability" of common nutrients can be greatly enhanced. As more beneficial nutrients reach the cells of the tissues and organs, which need them, greater health benefits are achieved.

Now that we've discussed the scientific reasons for administering nutritional substances intravenously, let's list and then briefly discuss each one. The antioxidant vitamins and minerals have all been named as beneficial cancer biotherapies and may improve immune functions. Beta-carotene, vitamins C, D, E, and K along with the antioxidant mineral selenium may be administered intravenously to spur the body's resistance to malignancy. Magnesium, calcium and molybdenum are also advocated by some practitioners and physicians.

Antioxidants, vitamins, and minerals may work individually and together in ways that are not well understood. Certainly, one of the leading theories given to explain beneficial results from these nutrients has to do with stimulation of the immune system. Immune system cells, such as macrophages, B cells and T cells may be increased in number and activity levels by antioxidant nutrients. Similarly, "cytokines," which are immune system boosters produced by the body, are increased by many of these nutrients. Cytokines, such as interferon, interleuken, and tumor necrosis factor are among those that may be stimulated by certain nutrients.

Antioxidants may also have direct effects on malignant cells. Some scientists believe they can "de-transform" cancerous cells and cause them to become normal, or at least to lose some of their deadly tendencies. Germanium and laetrile are often included in the intravenous therapies offered by alternative cancer physicians. They postulate that the internal environment of the body is made less conducive to the growth and survival of cancer cells by these substances.

Detoxification

Removal of accumulated metabolic wastes, poisons, and toxins may act to free the body's immune system and other functions to clear cancer cells away and promote their destruction. The body's physiological systems may be hampered by the presence of toxins resulting in their impaired ability to maintain normalcy and to maintain an environment unfriendly to abnormal cells such as cancer cells.

Hormone Balancing

Measuring, replacing and balancing the body's many hormones may result in a body that is stronger and more resistant to malignancy. Pregnenolone, DHEA, progesterone, estrogen, testosterone, androstenedione, growth hormone and melatonin may require specific balancing by a qualified physician to promote a healthy physiology. Caution must be taken as benefits and potential detriments of various ratios of these hormones must be considered for each individual and in relation to their disease and genetic background.

Homeopathic Remedies

Physicians familiar with the concept of "like treating like" have long advocated the use of homeopathic cancer medicines. Remedies such as Carcinosin may be chosen by the homeopathic physician in an attempt to spur the body's own energy to resist the cancer. Homeopathic theory advocates the use of homeopathic medicinals to change the "vital force" and thereby produce an internal physiology more conducive to health, and less conducive to the maintenance of malignancies.

Physical Therapy

Improving the functioning of the body via methods of physical therapy may also improve our resistance to neoplasms. Massage and drainage may improve our arterial flows of oxygen and nutrients to the tissues, aid in venous drainage, and facilitate lymphatic flow. Lymph, arterial blood and venous blood together make up the internal "transportation system" for delivery of nutrients and oxygen, and for the elimination of waste products, carbon dioxide and toxins. Physical therapy techniques may enhance this transportation system by directly improving flow and drainage patterns within our vascular system.

Mind and Spirit

It is instructive to note that many alternative practitioners have come to conclude that the spiritual and mental health of their patients seems to have a profound effect on their outcome. Sadly, much of modern conventional biomedical science seems to have forgotten these essential aspects of a human being's "being human" in favor of the sole use of interventional modalities such as drugs and surgery.

Any rational system of alternative therapy will necessarily need to encourage healthy self-discovery and the promulgation of a spiritual and emotional well-being within each person. Mental, emotional and spiritual well-being is directly tied to the physical well-being of each individual. As each discovers his or her own individual sense of spiritual and mental grounding, physical health will likewise be encouraged to flourish.

R-A Therapy in Nutrition

Many edible natural substances are known to cause re-differentiation and apoptosis. Re-differentiation connotes the ability of the body to cause cancer cells to revert to a normal, non-cancerous condition, while apoptosis means the potential for cancer cells to sense that they are abnormal and to then "kill" themselves. One of the most commonly available natural food substances that may cause R-A activity is lemon oil. Not only lemon oil, but the oils produced from the rind of all the citrus fruits contain a hefty dose of a health-giving natural chemical called "limonene."

Limonene has been shown to produce re-differentiation and apoptosis in many types of cancer cells. Each bottle of lemon oil contains almost 95% limonene by weight. So that ingesting lemon oil is tantamount to giving onself a nice dose of limonene. Lemon oil can be added to foods or drinks and ingested as a food. It is important to note that, as is true with all oils, lemon oil is most healthful if taken fresh. If it is allowed to spoil, it may not have the desired beneficial effects, and may instead cause harmful effects, such as allergic reactions.

Chapter Fourteen
INTEGRATIVE CANCER THERAPIES

Is it possible to achieve the best remission of cancer by combining the most powerful of the alternative cancer treatments with a sensible and conservative use of conventional oncology techniques? Many people have a tendency to choose medical care based upon their own individual philosophy about life, and their personal hopes and wishes as to what "should" be the best way to procede. Some, therefore reject all chemotherapy and radiation because they believe that there "should" be a more natural way to achieve cancer cure, while others immediately reject alternative care based upon the assumption that if regular doctors don't use it, then it cannot be of any value. However, in the final analysis, physicians must use whatever treatments work best, regardless of whether they are natural, or whether they are conventional orthodox medical treatments.

A growing number of physicians and their patients are choosing to combine the best of both worlds. They are seeking to design a program of cancer care that integrates the best of alternative medcine with the best and most sensible therapies available in conventional oncology.

Any rational combination of the two fields must rely upon choosing therapies from both categories that have definite promise to cause clinically significant benefit, and the lowest possible chance of harmful side-effects. Fortunately, the normal use of the alternative cancer therapies described in this book rarely lead to any harmful side-effects that are clinically significant. However, conventional treatments often do cause substantial negative side-effects, and therefore, a judicious use of them is essential to produce an integrative therapy program that is both effective and that has side-effects that are reasonably tolerable by the patient.

Fortunately, the positive beneficial side-effects of natural therapies often offset the harmful negative side-effects caused by the conventional treatments. When the two are combined, the negative effects of chemotherapy, radiation and even surgery are often dramatically lessened by the positive side-effects of the natural therapies. This is due to the fact that natural cancer-fighting substances often have manifold effects throughout the body, and these beneficial effects allow chemo-damaged nerves to repair themselves, increase the rate of wound healing following surgery, and allow the immune system to strengthen and rebuild iteself following radiation.

As a matter of fact, there are a multitude of ways in which many of the natural therapies strengthen, repair, enhance, and re-invigorate the tissues and physiology of the human body. It is fortunate that these beneficial side-effects of alternative therapies tend to negate the harmful effects of the leading conventional drugs, radiation, and surgical procedures.

Why consider using conventional and alternatives together? Why not just choose one or the other, and just request that type of care? The reason to combine the best of these two seemingly opposing methods of cancer care is that they tend to work synergistically. By synergistically, I mean that the beneficial, cancer-killing effects of both the natural and the more toxic conventional treatments often seem to combine to produce a more profound beneficial effect than would either alone.

It is widely accepted that alternative therapies are gentler and slower to act than are conventional treatments, but the alternative treatments, being more natural, cause less harmful side-effects. The natural therapies work in their own fairly unique ways to induce healing. The conventional treatments work by different mechanisms of action and in a more rapid and vigorous manner. Combining the various mechanisms of action of the alternative and conventional treatements hits the cancer from several angles simultaneously. Being hit from various angles of attack simultaneously makes it more likely that the cancer cells will respond by dying, or reverting to a non-cancerous condition, or both.

It also increases the odds that the entire tumor, or the entirety of the malignant cells in the body can be eliminated as a threat to the patient. There is also reason to surmise that the chance of any remaining cancer cells being able to mutate into a more resistant form is lessened by the judicious combination of natural and orthodox cancer care. In this chapter we will discuss the major conventional, orthodox methods, and then discuss rational ideas to integrate them with natural therapies.

Conventional Cancer Treatment

Surgery

The surgical removal of cancerous tumors is the oldest and most prevalent form of conventional therapy for the dreaded disease. Excision of as much of the cancerous growth as possible certainly makes common sense. This treatment works better for certain tumors than others, but is almost always the first step in the treatment of malignancies. Unfortunately, it is very rarely curative for most of the truly dangerous tumor types. Metastatic cancers, those that have spread out from their point of origin, are the most deadly and least amenable to cure by surgery alone.

This means that the majority of metastatic cancers which we fear the most, such as lung, kidney, colon, liver and brain cancers, are almost never cured by surgery alone since they are not usually discovered at early enough stages. Tumors that are discovered in their earliest stages, before eating their way into the deeper surrounding tissues or metastasizing to other organs, are the most amenable to cure by surgery. Examples of these include malignant polyps of the colon, certain skin cancers, and oral cancers discovered early.

The low cure rate from surgery alone leads conventional medical doctors to rely upon chemotherapy and radiation as follow-up treatments. These are intended to kill the residual cancer cells missed by the surgical excision of the main tumor mass and adjacent lymph nodes. In a minority of tumor types this triad of conventional therapies produces a decent cure rate. Unfortunately, most of the commonly feared cancers fail to be cured by the best efforts of physicians using these three powerfully destructive therapies.

Chemotherapy

Drugs that kill cells are used as chemotherapy agents. Known scientifically as "cytotoxins," (cyto=cell and toxin=poison), they are poisonous to the growing cells within our bodies. Oncologists who administer chemotherapy drugs to their cancer patients hope that the cancer cells will be preferentially killed by the drug. However, it is inevitable that beneficial cells of our body will also be killed or injured by the noxious agents. This loss or damage to beneficial cells leads to many of the harmful side effects of chemotherapy.

Although the ingestion of systemic poisons as a treatment for cancer seems to breach our common sense, it does sometimes work and at times makes sense scientifically. Certain tumors are preferentially killed off by chemotherapy drugs to such an extent that the peripheral damage to healthy normal cells is worth the trade-off. To those unfortunates who suffer through the side-effects of chemotherapy, the results of their normal cells being damaged in the "trade-off" can lead to much physical discomfort and debility. Therefore, it is imperative that doctor and patient alike carefully consider the likelihood of success against the suffering that will be inflicted through the use of chemotherapy drugs before beginning treatment with these cytotoxins.

Another reason for caution in the use of chemotherapy is the ill effect it may have on the immune system and upon the physiological functions of the body generally. Certainly, we must maintain a healthy immune system and good general health if we are to have the strength to "fight-off" the cancer. Many conventional physicians seem to conclude that the general state of health and the function of the immune system are of little importance in the fight against cancer. I believe that they are wrong. Improving the immune system and increasing the general state of health may be critically important for many cancer patients.

In my opinion, an integration of holistic techniques to strengthen the natural defenses against cancer should be undertaken concurrently with any conventional treatments chosen. When considering chemotherapy, it is crucial to weigh the probability of success against the damage that it will inflict upon the immune system and other healthy cells of the body.

Radiation Therapy

The use of radiation to destroy tumors is another of the primary treatment options commonly utilized in modern oncology. Radiation oncologists direct beams of gamma or other types of radiation at specific points within our body affected by cancer. This use of an "invisible knife" can destroy tumors in place. Another common use of radiation is to destroy the tissue surrounding a tumor after the tumor itself has been surgically removed.

When directed to a small, specific site, radiation therapy can effectively destroy a limited area and therefore spare other healthy tissues and organs from collateral damage. However, radiation directed to wider areas can harm important healthy tissues. Once again, the trade-off between risk and benefit must be carefully weighed. Patients are advised to question their doctors in detail as to the likelihood of success with a given treatment regimen as opposed to the possibility of damage to the immune system, damage to other healthy organs, and the physical suffering that may occur as the side-effects are manifested.

Certainly, conventional medical treatment of cancer may be worthwhile in specific cases. Hopefully, the billions of dollars currently spent on conventional cancer research will soon allow physicians to offer conventional therapies that have a much greater chance of success, and a greatly reduced side-effect profile than are presently available. Meanwhile, it seems advisable to seriously consider using the holistic approaches that are forgotten or forbidden by conventional medicine. These include the use of techniques to strengthen the immune system, detoxify the body, improve circulation of blood and lymph, restore oxygenation of the tissues, and allow the body to better utilize its own natural defenses against cancer.

Integrating Conventional and Alternative Treatments

Utilizing the best of both alternative and conventional care may yield more profound beneficial effects than using either alone. This is due to the fact that we can combine the various mechanisms of action of each type of care, allowing the cancer cells fewer avenues of escape, fewer ways in which they can evade being killed, less likelihood of mutating into a more difficult form, a diminished ability to metastasize, and less chance they can "hide" (remain in a physical location within the body, or within a tumor, which affords protection from the medicines) from the agents being used.

The leading options in alternative and conventional medicine for cancer may also work from slightly different versions of a similar mechanism of action. Once again, this may impart the benefits of hitting the cancer from slightly different angles, keeping the cancer from being able to easily change itself to evade the medicine. Some examples of this phenomenon are, 1) the anti-angioneogenesis agents, which are found in both alternative (squalene, convolvulus, and shark cartilage) and conventional medicine, and 2) primary pro-apoptotic agents, such as the conventional drug "aptosyn" and the alternative, natural substance "limonene."

It is very plausible that the pro-apoptotic actions of both aptosyn and limonene may work together in unison to produce a complementary and synergistic effect more profound than either alone would have produced. It also seems possible that shark cartilage and medical drugs that act to inhibit the development of new blood vessels to tumors can also work together, enhancing the chance each will work.

An Example of Integrative Therapy - Breast Cancer

Women diagnosed with breast cancer may well choose to undergo one or more of the conventional medical and surgical treatments for the disease after consultation with the doctor and carefully weighing the risks of the treatment to the benefits that can reasonably be expected. What additional therapies are advisable to strengthen her natural resistance to the malignancy? Let's briefly discuss this common malady and the natural treatments that may be integrated with conventional therapies.

Malignancies of the breasts are now occurring in what seems to be epidemic numbers. It is estimated that American women have a one in seven to ten chance of developing breast cancer. Why is this? At present, we do not know for sure. The consensus among scientists seems to be that multiple factors are predisposing American women to larger occurrences of this disease. Increased exposure to pollutants is one factor often cited. Heavy metals, such as mercury, lead and cadmium may play a role. Petrochemicals, solvents, pesticides, and a host of others are known to cause cancer in laboratory animals and may increase the incidence of breast cancer.

Another causative factor is the changing internal hormonal environment. Increased estrogen levels and estrogen activity seem to be occurring within the bodies of women who live in industrialized nations. Scientists believe that increased estrogen stimulation of the breast tissues may increase the likelihood that malignancies of the breast will develop. Estrogens are given to the animals that we eat in order to increase their growth rate. Ingestion of these food estrogens may chronically increase estrogen levels in women. Many polluting chemicals, particularly the pesticides, are also estrogenic in nature.

Changing reproductive habits may also, inadvertently, increase estrogen stimulation of the breast tissues. Statistics reveal that later age at first pregnancy and lowered rates of breast feeding increase the likelihood of later breast cancer. This may result from longer periods of estrogenic stimulation of the breast tissues, or a lack of counterbalancing hormonal influences by the hormones that control pregnancy and lactation. Estrogen dominance occurs when higher than normal levels of estrogen occur, or when estrogen is not balanced by corresponding levels of progesterone. Signs and symptoms of estrogen dominance are becoming more widespread and seem to correlate positively with the rising rate of malignancies of the breast.

What forms of biotherapies are available to combat breast cancer? How can physical bodies be strengthened to resist this dreaded disease? Recent experience by physicians, new scientific studies, and the experience of natural healers over the past hundreds of years point to quite a few such biotherapies. Let's begin with some common nutrients that have been found to influence breast cancer and can be administered intravenously to produce an even greater health benefit.

IV Nutritional Biotherapies

As with many cancers, certain antioxidant vitamins and minerals may be beneficial as breast cancer biotherapies. These include beta carotene, vitamins C, D, E, and K along with the antioxidant minerals magnesium, calcium, molybdenum and selenium, which can be administered intravenously to bolster a woman's natural resistance to breast cancer. Laetrile and germanium are mainstays of many alternative physicians in their intravenous therapies for breast cancer.

Food Nutrients

Algae, seaweed, medicinal mushrooms, fish oil, linseed oil, garlic, and fiber may all benefit the breast cancer patient. Fiber is a simple component of many foods, especially whole grains and vegetables. It can have a dramatically beneficial influence to halt cancer development or progression. In the case of breast cancer, these effects may occur through the moderating influence that dietary fiber can have on blood levels of estrogenic hormones, fats, and toxins. Dietary fiber binds these substances while they are in the digestive tract and causes them to be carried out of the body in the feces, rather than entering the bloodstream.

Algaes, seaweed, some mushrooms, and garlic may all affect breast cancer through unique natural chemicals that they possess. Garlic, for example, contains the natural chemicals allicin and allinin, which are both found to have many physiological effects that benefit the body. Beneficial species of mushrooms include the shitake and maitake, which have both been used for centuries by the herbal physicians of China and Japan to treat ailments including cancer. Lentinan, a derivative of the shitake mushroom, is also reportedly beneficial for breast cancer.

Fish oil and plant-derived flaxseed oil have also been reported as beneficial. Containing relatively high amounts of omega 3, these oils may help breast cancer cells to become more like normal cells. They may act to "de-transform" or inactivate cancer cells as they produce beneficial changes in the oily "lipid" outer membrane and inner nuclear membranes of the cancer cells.

Newer, More Powerful Biotherapies

The concepts of R-A therapy and telomerase treatments for cancer of the breast may prove the most powerful ideas in alternative medicine. Both concepts utilize the natural healing power of the cancer cells themselves, as they induce the genes that control each cancer cell to either revert the cell back to a normal, non-malignant condition, induce it to kill iteself, or allow the cell to age like normal cells do, thus allowing them only a limited reproduction life-span.

Other Biotherapies

A brief list of other biotherapies that may be beneficial in breast cancer includes certain herbs, canthaxantin, cesium and rubidium, gossypol, Coleys Toxins, heat therapy, hydrazine, indoles, iscador, megace, MTH-68, super oxide dismutase supplements, staphage lysate, suramin, tamoxifen, benzaldehyde (a breakdown product of amigdalin) and thioproline.

Integrating Conventional Modalities

We may choose to utilize many different orthodox treatments in a case of breast cancer. Review of the case may show that surgical excision of a lump or tumor within the breast tissue makes sense. We would first prepare the patient by administering natural therapies intended to weaken and shrink the tumor. We want the cancer to be as weak as possible at the time it is surgically removed from the body. This is premised upon the notion that the cancer may be less likely to metastasize, and any remaining cancer cells may have less ability to mutate into a more vigorous form if they are weakened by pre-surgical treatment with natural substances.

The pre-surgical therapy is also intended to strengthen the immune system. A strong immune system is highly prized in alternative medicine, since it is assumed that the immune system is essential to good health and recovery from surgery, or any other physical stressor.

Similarly, following radiation or chemotherapy in a breast cancer case, we would want to re-invigorate the immune system. But why would any patient consider toxic methods like chemotherapy and radiation if they can use alternative therapies? Primarily to seek the best attributes of both natural and conventional modalities. Radiation or chemo often seem to work faster, better and more thoroughly when combined with alternative modalities. Another beneficial synergism is that chemotherapy and radiation treatments seem to cause less adverse side-effects when natural therapies are administered concurrently.

Likewise, the slower, gentler anti-cancer effects of alternative means may be enhanced when the faster and more abrupt effects of the conventional therapies are administered simultaneously. Certainly, integrating the best of alternative and conventional breast cancer care is an interesting and compelling option that many women are choosing.

Highlights of Integrative Cancer Care Options

The concept of integrating the best modalities of both orthodox and natural cancer modalities rests upon two premises. These are, 1) synergism, and 2) the amelioration of the negative side-effects of conventional medical treatments.

Synergism may occur as the differing mechanisms of conventional and natural care modalities work together to produce a more profoundly beneficial effect than either alone would have done. Amelioration of the harmful side-effects of chemotherapy, radiation and surgery occurs as the healing properties of the natural substances used in alternative medicine act throughout the body to reverse damage the conventional treatment did to essential healthy cells, organs, and to the immune system.

The most promising ways to achieve beneficial synergism and the amelioration of negative conventional care side-effects are to induce apoptosis, use telomerase treatment to weaken the cancer cells genetically, and to induce re-differentiation, the reversion of cancer cells back into a non-cancerous state.

Apoptosis, a form of pre-programmed cell suicide, is thought to play an essential role in both cancer and aging. Premature apoptotic loss of essential cells within vital organs may be causative in the general aging process, in heart disease, and in other aging-related degenerative diseases. Conversely, the loss of normal apoptotic mechanisms may allow malignant cells to proliferate without being eliminated through organism-protective programmed cell suicide.

The natural substances of R-A therapy seek to induce apoptosis in a gentle and non-toxic manner by encouraging the genes of apoptosis within the cancer cells to turn on, and restore their normal functions, which would be to cause the cancer cell to enter a program of suicide, thus eliminating the cancer cells from the body.

Radiation and chemotherapy may also play a role in causing cells to enter apoptosis, but via a different mechanism. The conventional modalities seem to cause apoptosis by being so toxic to the cancer cells that they are induced to die, perhaps via apoptosis. This of course depends upon the apoptotic genetic mechanisms being in place and active. If they are not available for the toxic actions of radiation or chemotherapy to induce cell death via apoptosis, then the cells may instead continue living, and be induced instead to mutate into more radical forms that will later be more difficult to kill or eliminate.

I speculate that one of the primary benefits of the natural substances utilized in alternative cancer medicine is to maintain or re-establish an active and vigorous apoptotic pathway. It also seems plausible that if this pathway is not active, the effects of radiation and chemotherapy may be to produce more severely deranged malignant cells rather than killing them. This defect in conventional cancer care, namely the inability to kill cancer cells and to cause them to become more malignant if apoptosis is not induced, may explain why it is so difficult to actually cure cancer via the use of toxic orthodox treatments.

Book Conclusion

The purpose of this book has been to explain and elucidate the ways in which natural healing can be induced, and methods by which health and longevity can be maintained. Many powerful natural therapies for aging, cancer, heart disease, and the degenerative diseases of aging are now available. These are often used alone, but may also rationally be utilized in conjunction with sensible conventional medicines and treatments.

R-A therapy seeks to induce cancer cells to both kill themselves through apoptosis and to cause remaining cancer cells to revert to a non-cancerous condition. Telomerase treatments seek to genetically alter cells in benefical ways. It is intended to cause cancer cells to become genetically weaker, so that they can better be eliminated from the body by natural and/or conventional means. It is intended to cause our healthy, normal cells to resist the aging process.

The combination of R-A therapy, telomerase treatments and all the other varied natural and conventional therapies described in this book offers great hope to all of us who wish to live long, healthy, active lives, free from disease!

Appendix
APOPTOSIS

This chapter consists of the thesis I submitted while a masters degree student at Stanford University. This thesis (actually it was a research project in lieu of an actual "thesis project") was written based upon the laboratory research performed by myself and undergraduate students in the lab. It is presented here as further information for those interested in a more detailed scientific treatise on the topic.

Variable "Apoptotic Fuse" Timing Influences Cultured Mammalian Cell Death To Colcemid, Aphidicolin And Cisplatin

Abstract

Unique "apoptotic fuse" mechanisms in various human cancer cell lines may produce differences in propensity to apoptose and in time to apoptosis when the cells are exposed to cytotoxic agents. The "apoptotic fuse" represents each cells' intrinsic pathway of sensing cell cycle perturbations and effectuating an apoptotic response. Cytotoxicity to continuous exposure of cultured mammalian cells to aphidicolin, a DNA synthesis inhibitor, colcemid, a disrupter of the mitotic spindle apparatus, and cisplatin, a DNA cross-linking agent, results from the activation of each cell line's unique "apoptotic fuse" mechanism. Apoptotic cell death is indicated in this study by, 1) flow cytometry to detect chromatin condensation and alterations in cell size, 2) detection of DNA degradation via the generation of nucleosome length DNA fragmentation (DNA ladders), and 3) vital dye exclusion.

The time required from initial exposure to cytotoxic drugs until the initiation of apoptosis in three human cancer cell lines (Hela-S3 fibroblast derived, CEM T-cell lymphoma, and Tera-2 embryonic carcinoma) varies strikingly, and can range from four to forty-eight hours. The cell lines studied each have unique growth properties and divergent embryonic origins, which may have endowed them with differing "apoptotic fuses" and mitotic control mechanisms, resulting in variations in time to apoptosis. Cell lines responding with rapid apoptosis are those derived from embryonic (Tera 2) and immune system (CEM) origins, whereas cells derived from human fibroblast origins (HeLa S3) exhibit a longer "apoptotic fuse time." We postulate that the embryonic and immune system cells exhibit a greater propensity to apoptose due to their inherent need to select unwanted lineages during clonal selection or embryonic differentiation.

Colcemid treated HeLa cells are found to apoptose less readily than CEM cells, which apoptose less readily than Tera-2. The varying times to apoptosis reflected in these differing "apoptotic fuses" may reflect inherited apoptotic propensities which may commonly remain intact after malignant transformation of the cell lineages, resulting in different "apoptotic fuse" times to "detonation," i.e. pre-programmed cell death.

The observed variations in the timing of mitotic arrest and apoptosis among cells from differing tissue origins suggests that each cell type responds uniquely to cell cycle perturbations via intrinsic "apoptotic fuse" mechanisms. These differences in reactions to cytotoxic agents lead to the conclusion that the success or failure of clinical chemotherapeutic treatment regimens hinge on inherent differences in apoptotic responses of the particular cell type, and especially of the particular tumor cell clone, have been predetermined experimentally. Whenever possible during clinical treatment with chemotherapy, concurrent measurements of the apoptotic response to treatment through flow cytometry or other available means may prove beneficial in determining efficacious dosages and the most advantageous timing of their administration.

Introduction

It has been postulated that various human cancer cell lines possess inherent differences in their predilection to undergo apoptosis (Kung, Schimke 1990). Such differences in the inherent propensities of various cell lines to apoptose may be due to unique mechanisms that sense perturbations in cell cycle progression and then effectuate an apoptotic response to the perturbation. The "apoptotic fuse" concept represents each cell's intrinsic pathway of sensing cell cycle perturbations and effectuating an apoptotic response.

As research to define the apoptotic process advances, it is becoming evident that most cell lines hold significant portions of the apoptotic pathway in common. The apparent prevalence of these common "fuse" components may lead to the conclusion that cells of different lines should apoptose in a similar manner and within a similar time frame. However, our research findings tend to dispute that assumption. We find that various cell lines present differing propensities to apoptose and do so within different time periods.

Disruption of cellular replication or homeostatis caused by chemotherapeutic agents results in an active process of cellular self-destruction that has features distinct from those found in degenerative or necrotic pathological cell death (Kung, Zetterburg, 1990). Evidence from biochemical and morphological analysis indicates that apoptosis is an active, directed process of cellular "suicide" in response to significant perturbations in the cell cycle for homeostatic functions. The biochemical elucidation of the apoptotic pathway has been the focus of much recent research.

This research indicates that although the apoptotic pathway and its initial triggering mechanism may differ from one cell line to another, certain components of the apoptotic pathway, such as p53 gene expression, seem to be held in common (Holbrook and Fornace, 1991; Fornace, 1992). Expression of the BC12 gene, the Retinoblastoma gene, and the Ataxia Telangiectasia gene have also been found to play an important role in modulating the apoptotic pathway. It seems likely that other positive and negative affectuators of a common apoptotic pathway will be discovered and described in the near future, eventually leading to a complete elucidation of the primary components of the "apoptotic fuse." Differing expressions of these "fuse" components, or extrinsic modulators such as growth factors, may prove the basis of the differences seen in time to apoptosis among various cell lines.

The pre-programmed sequence of events leading to cellular suicide may be described as unique "apoptotic fuse" mechanisms if it can be shown that, 1) cell lines differ in their propensity to enter the apoptotic process as a response to inhibition of cell cycle progression induced by cytotoxic agents, 2) cell lines differ in their rate of undergoing apoptosis, and 3) various cell lines respond in unique ways to certain cytotoxic agents or survival factors.

Several recent discoveries have aided the understanding of the directed, pre-programmed suicide process that constitutes the "apoptotic fuse." Efforts to understand the pathways involved in cell cycle checkpoints, cell cycle arrest, and apoptosis have resulted in the implication of the p53 gene in these processes (Hartwell, 1989). First thought to be an oncogene, p53 is now understood to act as a tumor-suppressor gene. Its proteinacious transcript appears to act as an inhibitor of cell cycle progression at G1/S, possibly by acting as the transcription agent for the apoptotic p21 protein, which subsequently blocks Cdk enzymes thus halting cellular division (Marx, 1993).

Genes currently implicated in the apoptotic response include Bcl2 which may act to modulate p53 activity by decreasing p53 transcription or the phosphorylation of the p53 protein. The Retinoblastoma gene also seems to act as a cell cycle regulating tumor suppressor gene. Similarly, the Ataxia Telangiectasia gene is a possible upstream regulator of p53. Other genes directing the suicide program's "apoptotic fuse" may well be discovered as the result of continuing research into the mechanisms of programmed cell death.

Chemotherapy depends in great measure upon the active transition of cells through the cell cycle. Active passage of cells through the cell cycle appears to be an important factor aiding in the "lighting" of the apoptotic fuses "wick." Lack of proliferative and growth activities therefore makes it less likely that cytotoxic agents such as cisplatin, aphidicolin and colcemid will result in apoptotic cell death.

Multiple scientific papers have given evidence that growth or "survival" factors such as Platelette Derived Growth Factor (PDGF) and Insulin-like growth Factor-1 (IGF1) can act to initiate and maintain cell cycle progression in previously quiescent cells (Pledger, 1977) (Pledger, 1978) (Trojan, 1993). PDGF has been shown to cause the quiescent cells to enter G1 from G0, and IGF1 has been shown to assist the continued progression of the cells to complete a cycle (Pledger, 1977). PDGF and IGF have therefore been termed "competence" factors and "survival" or "progression" factors, respectively, by many authors. It appears that these competence and survival factors dramatically influence the initiation sequence of apoptosis, particularly in certain cell types. These factors may therefore approve to be important modulators in the running each cell's "apoptotic fuse" program.

Materials and Methods

This segment is deleted in this publication.

Results

Experimental results measuring apoptotic responses in three cell lines were obtained through flow cytometry indications of apoptotic DNA "slideback," microscopic examination of Hoechst dye stained DNA, propidium iodie exclusion, and agarose gel electrophoresis for presence of "DNA ladders." Results from these various analyses as presented here reveal that the time required from initial exposure to cytotoxic drugs until the initiation of apoptosis in the three human cancer cell lines studied (HeLa-S3) fibroblast derived, CEM T-cell lymphoma, and Tera-2 embryonic carcinoma) varies strikingly, and can range from four to forty-eight hours.

These findings may reflect the unique growth properties and divergent embryonic origins of the cell lines studied, with embryonic and immune system cells exhibiting a greater propensity to apoptose. The three cell lines may have been endowed with differing "apoptotic fuse" and mitotic control mechanisms, resulting in variations in time to apoptosis. The cell lines evidencing the most rapid apoptosis are those derived from embryonic (Tera2) and immune system (CEM) origins, whereas cells derived from human fibroblast origins (HeLa S3) exhibit a longer time to apoptosis.

Colcemid treated HeLA cells arrest readily in mitosis but apoptose less readily. CEM cells apoptose less readily than do Tera2 cells, and apoptose more readily than HeLa. The varying times to apoptosis may reflect different inherited apoptotic propensities, which may commonly remain intact after malignant transformation of the cell lineages, resulting in different time periods needed to achieve preprogrammed cell death. The inherent need for rapid apoptosis in embryonic cells may derive from their ability to select unwanted lineages during differentiation, while the immune system derived CEM cells may similarly select out unwanted clones via a vigorous apoptotic system.

COLCEMID: Flow cytometry was employed to measure the differences in apoptotic responses among the Tera-2, HeLaS3, and CEM cell lines when continuously exposed to various concentrations of colcemid, aphidicolin, and for all but CEM, cisplatin. When treated with colcemid, each cell line showed signs of a G2/M arrest followed by a slideback in the form of a subpopulation of G2 cells (which would normally have 4C DNA) that are smaller and appear to have less than 4C DNA. Histograms obtained of Tera-2 cells, harvested at four hour intervals, after continuous exposure to a completely inhibitory concentration of colcemid (70ng/ml) were obtained. The cells showed indications of 2 accumulation with a slight decrease in cell size in the 4C DNA population by four hours, a well-defined subpopulation of cells with less than 4C DDNA by eight hours. At eight hours we also observed indication of a subpopulation of less than 2C DNA cells. The subpopulation of less than 2C DNA cells may represent micro cells produced after an aberrant round of mitosis, as well as cells that have died after progression through mitosis. As time progressed, the G1 peak disappeared, as the cells progressed through mitosis and apoptosis.

Histograms of CEM cell populations fixed at four-hour intervals during continuous exposure to colcemid (70ng/ml) revealed indications of apoptosis seen at 16 hours with cell shrinkage of 4C cells after accumulation at the G2/M stage of the cell cycle. By 24 hours, the slideback is well defined and the number of cells remaining in G2 continues to decrease. Cells containing less than 2C DNA are also becoming apparent. These 2C cells may also be due to micro cell formation as well as cells which have died after undergoing mitosis. We observe a G1 population at 36 hours, which may represent cells that never go through the cell cycle or cells which perhaps progress through mitosis without undergoing apoptosis. We note that CEM cells never exhibit a great degree of cell shrinkage in G2 due to the paucity of cytoplasm commonly present in B and T cells.

Histograms of HeLa S3 cells fixed at four-hour intervals during continuous exposure to colcemid (70ng/ml) were obtained. Contrasting with the previous two cell lines, HeLa S3 cells exhibit a total arrest at 4C DNA content by 24 hours, prior to any sign of apoptotic slideback. Analysis reveals initial stages of cell shrinkage and slideback from G2 at 32 hours. By 40 hours, the apoptotic population is well defined.

Microscopy revealed apoptotic changes in cells stained with Hoechst DNA stain and propidium iodide. Aberrant chromosome condensation was revealed upon microscopic examination of the specimens while propidium iodide exclusion as shown under ultraviolet light was used to indicate cell viability, since cells with intact membranes exclude propidium iodide. Cells were counted as apoptotic if they were impermeable to propidium iodide but allowed aberrant chromosome condensation.

Graphic representations of the cell counts of normal, dead, and apoptotic populations as determined by microscopic examination at four-hour intervals were developed. The control Tera2 counts show a high percentage of apoptotic and dead cells, a common finding in this naturally unstable cell line. This inherent instability and propensity to spontaneous death was seen with Tera2 throughout the experiments. Analysis showed an increase in the percentage of dead and apoptotic cells as early as eight hours in Tera2, whereas this is seen at twenty hours in CEM, and at thirty-six hours in HeLa S3 cells.

By 36-40 hours, all three cell lines exhibited similar percentages of dead and apoptotic cells when subjected to colcemid. The time points at which the initial increase in numbers of apoptotic and dead cells were observed correlates closely with the times at which slideback from 2 is observed in the DNA histograms obtained by flow cytometry. This correlation indicates that chromosome condensation and slideback occur at comparable times. The foregoing findings were further confirmed by detection of chromosomal DNA degradation with agarose gel electrophoresis. Ethidium bromide stained gels revealed DNA "ladders" by 24 hours in CEM, 32 hours in HeLa S3, while Tera2 failed to show clear indications of degradation.

Results of cell counts, electrophoresis and flow cytometry indicate that Tera2 cells apoptose earlier than do CEM and HeLa S3 cells to colcemid exposure, while CEM responds with a more rapid apoptosis than do HeLaS3. The results also show that all three lines have undergone a significant percentage of apoptosis, out of 2/M, by 40 hours.

APHIDICOLIN: Analysis of the three cell lines by flow cytometry shows that a "slideback" phenomenon occurs out of the G1/S cell cycle peaks. This "slideback" phenomenon is caused by apoptopic DNA degradation and chromatin condensation. Flow cytometry analysis of Tera2 and CEM cells reveals the beginning of measurable apoptosis at 12 hours with indications of more significant apoptosis at 24 hours. HeLa S3 cells reacted differently, evidencing apoptosis at 24 hours. Microscopic counts correlating percentages of normal, dead and apoptotic cells at four hour intervals correlate with the flow cytometry data. The charting of these microscopically determined cell counts reveals a dramatic rise in apoptotic cells after 24 hours for HeLa S3, and a small rise in apoptotics at 12 hours followed by a more significant increase at 24 hours in both CEM and Tera2.

The results of cell counts and flow cytometry indicate that Tera2 and CEM cells respond with a more rapid apoptosis (12 hours), than do HeLa S3 (24 hours). All three lines were shown to have undergone a significant percentage of apoptosis, out of G1/S, by 24 hours.

CISPLATIN: Flow cytometry and microscopic scoring data of the relative percentages of live, apoptotic and dead cells show that both Tera2 and HeLa S3 die out of G1/S via apoptosis when subjected to continuous cisplatin treatment. Tera2 were found to arrest and die out of G1/S through apoptosis more rapidly (8 hours) and at lower concentrations of cisplatin (200ng/ml) than did the HeLa S3 cells. HeLa S3 cells required 24 hours to attain significant G1/S arrest at a Cisplatin dosage concentration of 240 ng/ml, and lagged Tera2 by requiring 48 hours to exhibit significant apoptosis.

Discussion

Cytotoxicity to exposure of cultured mammalian cancer cells to the chemotherapeutic agents used in this study, including aphidicolin, a DNA synthesis inhibitor, colcemid, a disrupter of the mitotic spindle apparatus, and cisplatin, a DNA cross-linking agent, resulted from the activation of the unique "apoptotic fuse" mechanism inherent in each cell line tested. Resultant from this process of pre-programmed cellular suicide was the variability in time to apoptosis seen for the three cell lines. In these experiments, HeLa, CEM, and Tera2 cells were studied after continuous short term exposure to Colcemid, which tends to arrest cell cycle progression in motosis, and Aphidicolin, which inhibits cell progression through S phase. Colcemid acts via tubulin inhibition, which results in disruption of the mitotic spindle apparatus. Aphidicolin is a DNA synthesis inhibitor causing derangement of the normal S phase replication of chromosomes. Both HeLa and Tera2 were studied after continuous short term exposure to Cisplatin (a heavy metal complex consisting of a platinum atom surrounded by cis-positioned chloride atoms and ammonia molecules), which acts to cause DNA cross-linking resulting in inhibition of DNA synthesis, chromosomal replication, and cellular homeostatic functions (Barry, 1990). These experimental results suggest that HeLa cells arrest readily in mitosis but apoptose less readily than either CEM or Tera2, and that CEM cells apoptose more readily than Tera2.

The inherent differences reflected in these results might reflect the cell line's individual propensities for apoptosis as determined by evolutionary selection. As an example, it seems reasonable to expect a more rapid burning of the apoptotic fuse in the CEM line since they are deriving from lymphocyte lineages which have a significant need to select certain types of unneeded cells through apoptosis during the clonal selection process. On the other hand, fibroblast derived cell lines such as HeLa have less of a need for rapid apoptosis in the normal course of functioning as components of relatively stable, somatic tissue. The inherited apoptotic propensities may therefore commonly remain intact after malignant transformation of the cell lineages, resulting in different "apoptotic fuse" times to "detonation," i.e., cell death.

The purpose of the clinical administration of cytotoxic chemotherapeutic agents is to "kill" cancer cells. However, various cancer cell lines respond in vastly different ways to these agents, in many instances causing reduced efficacy of treatment. Resistance mechanisms as studied in the laboratory classically involve the exposure of cells to high concentrations of an agent for up to two to four weeks at which time surviving colonies are analyzed. Under these conditions cell death results from apoptosis.

Resistance mechanisms characteristically involve alteration in cells so that the action of the agent is prevented, such as amplification of target enzyme, increased drug efflux, or point mutations altering drug effect. These selection protocols dictate that any surviving cells must prevent the activity of the cytotoxic agent. Long term exposure protocols to define drug resistance are problematic in that they do not correspond to normal clinical usage of the agents. Normal clinical usage of the agents produces a much shorter and effective drug exposure time (24 hours maximal), as compared to the long term exposure protocols.

This laboratory has previously proposed that cell killing related to short-term exposure to cytotoxic agents can depend upon how cells respond to a state of prior inhibition (Schike et. al, 1991). This laboratory has also shown that apoptosis occurs in CHO and

HeLa S3 cells only after a state of suspended cell cycle progression equivalent to approximately 30-40 hours, and that the time from cell cycle arrest to onset of apoptosis is not constant for all cell cycles.

Some cell lines such as Tera-2, SB, and CEM initiate apoptosis much sooner following colemid-induced mitotic arrest than do HeLa, VA-13 or CHO cells. Additionally, expression of Bcl-2 can prolong the time to onset of apoptosis in mitotically synchronized HeLaS3 cells from 30-36 hours post mitosis to approximately 60-72 hours (S. Sherwood, unpublished). Thus, if the time to apoptosis is as variable among cancer cells as it appears to be among various other cell lines, then a limited time of drug exposure as occurs clinically may well fall short of the required apoptotic fuse time for the cancer cell line being treated, although it may have been adequate for other cancers.

Therefore, the use of "aggressive" treatment protocols with high bolus drug concentrations may be effective by prolonging time of effective concentration rather than solely by the effects of the peak concentration of the drug reached at the tumor site. Awareness of each cell line's intrinsic "apoptotic fuse time" also points to the benefits potentially derived from following the cells by flow cytometry, whenever possible, during the clinical course of treatment. Following the tumor cells via flow cytometry during treatment may also aid in the detection of a new clone, which may have a different apoptotic fuse time than did its predecessor.

Another factor potentially affecting clinical treatment of cancer are our findings that suggest that if cells are subjected to drug concentrations for less than the length of time required to induce apoptosis, or at a concentration that only partially inhibits cell cycle progression, cell death occurs by the generation of microcells, with tri or quadripolar mitoses. This occurs with agents that disrupt mitotic spindle formation and with inhibitors of DNA synthesis.

Cell killing, with low concentrations of the DNA cross-linking agent Cisplatin, also involves aberrant mitoses and the generation of micro-cells (J. Sheridan, to be published). These results are suggestive of a disruption of a putative "centrosome cycle." Normally the centrosome is replicated in G1, but takes up residence at the opposite pole only in G2. When employing low drug concentrations that slow aspects of cell cycle progression, mitosis is often delayed. If centrosome replication occurs in the linear-frame of a normal cell cycle, centrosome duplication may occur prior to a delayed mitosis resulting in multi-polar mitoses. These multi-polar mitoses set off the "apoptotic fuse," resulting in apoptotic cell death. Kuriyama and Borisy have indicated that aspects of centrosome replication occur in the absence of DNA replication (Kuriyama, Borisy, 1981). One might therefore suggest that a putative "centrosome cycle" would have a feed-back control mechanism to assure that chromosome replication does not precede mitosis under conditions of drug induced, cell cycle delaying perturbations.

These suggested feed-back control mechanisms may be found to be analogous to the multiple cell cycle checkpoint controls which influence mitosis (Murray, 1992) (Hartwell, 1989). Several observations suggest such control. First, certain cell lines studied by this laboratory (VA-13), do not undergo microcell formation when treated with low concentrations of colcemid, although they are clearly responding to the colcemid with mitotic delay. The finding that not all cell lines undergo abnormal mitoses with resultant microcell formation suggests that they may have some type of intrinsic control over the centrosome cycle. Additionally, cancer cells have been reported to be more prone to aberrant mitoses (tri and quadripolar mitoses) than are normal cells (Levine, 1991). This leads us to suggest that a critical determinant of the success or failure of treatment may involve whether or not the cells undergo aberrant mitoses when subjected to partially inhibitory or short term, high concentration drug exposure as occurs during clinical treatment regimes.

The concepts of an "apoptotic fuse" unique to various cell lines, lends itself well to the understanding of chemotherapeutically induced programmed cell death and resistance mechanisms to chemotherapy. The observed variations in the timing of mitotic arrest and apoptosis among cells from differing tissue origins suggests that each cell type responds uniquely to cell cycle perturbations via intrinsic "apoptotic fuse" mechanisms.

These differences in reactions to cytotoxic agents lead to the conclusion that the success or failure of clinical chemotherapeutic treatment regimens hinge on inherent difference in apoptotic response times to a state of inhibition within the time frame of treatment. Therefore, the clinician can expect a greater likelihood of success when the apoptotic responses of the particular cell type, and especially of the particular tumor cell clone have been pre-determined experimentally. One of the most difficult problems for the clinical practice of oncology is to maintain therapeutic tissue and plasma levels of drug for a time period sufficient to inhibit the activities of proliferative neoplastic cells and thereby induce apoptotic demise via perturbations of the cell cycle check point systems. Whenever possible during clinical treatment with chemotherapy, concurrent measurements of the apoptotic response to treatment through flow cytometry or other available means may prove beneficial in determining efficacious dosages and the most advantageous timing of their administration. Elucidation of each cell line's "apoptotic fuse" will enable clinicians to deliver the proper concentration of drug, and to maintain therapeutic levels for the time required for the "fuse" to "detonate" the cancer cell via a pre-programmed demise.

The "apoptotic fuse" concept also engenders a new type of resistance mechanism. This resistance mechanism involves variable responses of cells to inhibitory concentrations of cytotoxic agents rather than resistance mechanisms involving prevention of drug action (amplification of target enzyme, increased drug efflux, or point mutations altering drug effect).

Further studies to determine the actual type of cell killing mechanisms and chemotherapy resistance mechanisms may well show that the processes of the "apoptotic fuse" are of great clinical significance. Monitoring the actual mechanisms of chemotherapeutically induced cell killing during the clinical course of treatment may point the clinician to the most efficacious regimen for treatment. A more in depth future understanding of the mechanisms of the "apoptotic fuse" may also enable novel interventions in the problems of senescence and immune system diseases including AIDS.

References

Holbrook, N. J., Fornace, A. J., Jr., "Response To Adversity: Molecular Control Of Gene Activation Following Enotoxic Stress," *New Biologist* 3(9) (September 1991): 825-33.

Fornace, A. J., Jr., "Mammalian Genes Induces By Radiation: Activation Of Genes Associated With Growth Control," *Annual Review of Genetics* 26 (1992): 507-26.

Pledger, W. J., Stiles, C. D., Antoniades, H. N., and Scher, C. D., "An Ordered Sequence Of Events Is Required Before BALB/C-3T3 Cells Become Committed To DNA Synthesis," *Proceedings of the National Academy of Science USA*, Vol. 76, No. 6 (1977): 2839-43.

Marx, Jean, "How p53 Suppresses Cell Growth," *Science*, Vol. 252 (1993): 1644-45.

Trojan, J., Johnson, T. R., Rudin, S. D., Ilan, M. L., Tykocinski, J. Ilan, "Treatment And Prevention Of Rat Gliobalstoma By Immunogenic C6 Cells Expressing Antisense Insulin-Like Growth Factor 1 RNA," *Science*, Vol. 259 (1993): 94-97.

Kung, A. L. Sherwood, S. W., and R. T. Schimke, "Cell Line Specific Differences In The Control Of Cell Cycle Progression In The Absence Of Mitosis," *Proceedings of the National Academy of Science* (USA) 87 (1990): 9553-57.

Schimke, R. T., Kung, A. L., Rush, D. R., Sherwood, S. W., "Differences In Mitotic Control Among Mammalian Cells, The Cell Cycle." *Cold Springs Harbor Symposium on Quantitative Biology*, Vol. LVI (1991): 417-25.

Hartwell, L. H. and T. A. Weinert, "Checkpoints: Controls That Insure The Order Of Cell Cycle Events," *Science*, Vol. 246 (1989): 629-34.

Kung, A. L., Zetterburg, A., Sherwood, S. W., and R. T. Schimke, "Cytotoxic Effects Of Cell Cycle Phase Specific Agents: Result Of Cell Cycle Perturbation," *Cancer Research* 50 (1990): 7307-17.

Barry, M. A., Behnke, C. A., and Alan Eastman, "Activation Of Programmed Cell Death (Apoptosis) By Cisplatin, Other Anticancer Drugs, Toxins, And Hypothermia," *Biochemical Pharmacology*, 40(10) (1960): 2343-62.

Ross, R., Raines, E, and Bowen-Pope, D. F., "The Biology Of Platelet-Derived Growth Factor," *CELL*, Vol. 46 (1986): 155-69.

Siegfried, Z., and Ziff, E. B., "Transcription Activation B Serum, PDGF, And TPA Through The C-Fos DSE: Cell Type Specific Requirements For Induction," *Oncogene* 4 (1989): 3-11.

Murray, A. W., "Creative Blocks: Cell Cycle Checkpoints And Feedback Controls," *Nature* (1992): 359, 599.

Kuriyama, K. and Borisy, G. C., "Centriole Cell In Chinese Hamster Ovary Cells As Determined By Whole-Mount Electron Microscopy," *Journal of Cell Biology* 91 (1981): 814-821.

Levine, Douglas S., Sanchez, Carisa A., Rabinovich, Peter S., and Reid, Brian J., "Formation Of The Tetraploid Intermediate Is Associated With The Development Of Cells With More Than Four Centrioles In The Elastase-Simian Virus 40 Tumor Antigen Transgenis Mouse Model Of Pancreatic Cancer," *Proceedings of the National Academy of Sciences,* USA 88(15) (August 1991): 427-31.